Interdisciplinary Contributions to Archaeology

Archaeology stands alone among the sciences in its attempt to enlighten us about the entire record of humankind. To cover such a broad range of time and space, archaeologists must ensure that their findings are integrated into broader spheres of scientific knowledge. The IDCA series aims to highlight the collaborative and interdisciplinary nature of contemporary archaeological research.

Topics the series has covered include:

- Paleoecology
- Archaeological Landscapes
- Statistical Approaches
- Laboratory Methods
- Human Biological and Cultural Evolution
- Human Nutrition
- Emergence of Agriculture and Pastoralism

For a copy of the proposal form, please contact Christi Lue (christi.lue@springer.com). Initial proposals can be sent to the Series Editor, Jelmer Eerkens (jweerkens@ucdavis.edu). Proposals should include:

- A short synopsis of the work or the introduction chapter
- The proposed Table of Contents
- The CV of the lead author(s)
- If available: one sample chapter

We aim to make a first decision within 1 month of submission. In case of a positive first decision the work will be provisionally contracted: the final decision about publication will depend upon the result of the anonymous peer review of the complete manuscript. We aim to have the complete work peer-reviewed within 3 months of submission.

This book series is indexed in SCOPUS.

For more information, please contact the Series Editor at (jweerkens@ucdavis.edu).

Karenleigh A. Overmann

Cultural Number Systems

A Sourcebook

Springer

Karenleigh A. Overmann
Center for Cognitive Archaeology
University of Colorado, Colorado Springs
Colorado Springs, CO, USA

ISSN 1568-2722 ISSN 2730-6984 (electronic)
Interdisciplinary Contributions to Archaeology
ISBN 978-3-031-83382-3 ISBN 978-3-031-83383-0 (eBook)
https://doi.org/10.1007/978-3-031-83383-0

This Springer imprint is published by the registered company Springer Nature Switzerland AG
The registered company address is: Gewerbestrasse 11, 6330 Cham, Switzerland

Acknowledgments

I'd like to thank the many friends and colleagues who have helped me in this particular research. First are those who have lent me their time and expertise in languages both modern and ancient: Mohammad "Parsa" Daneshmand (Arabic), Antonis Iliopoulos (Greek), GUO Jue (Chinese), Annick Payne (German and Greek), and Alison Samuels (Latin). Their help in negotiating arcane dialects in ancient texts and unreadable fonts (Fraktur, Gothic, and Blackletter, I'm looking at you) was invaluable. Next are Don Pawl, the Interlibrary Loan coordinator who has spent countless hours tracking down obscure references for me, and Chongsook Son, the friend who brought several useful articles on Korean numbers to my attention (she also mentioned that Google Translate now works on images, allowing me to deal with scripts like Han'gŭl that aren't easily retyped on a QWERTY keyboard). Jean-Paul Potet graciously answered my questions about Old Tagalog. Kenichi Takashima kindly provided me with information on the oracle bone numbers. Two reviewers provided constructively critical feedback that greatly improved the manuscript; any remaining mistakes are completely mine. Of course, I remain profoundly grateful to the scholars who mentored my education and who continue to inspire my research: Fred Coolidge, Tom Wynn, and Lambros Malafouris. Last and never least are my grandfather, Carl William Stewart, who gave me my start in life; my children, Archie and Barbara, and their families; Will, the son we lost in 2016; and Bill, my best friend, anchor, soulmate, and husband of 44 years.

Contents

List of Figures

List of Tables

Chapter 1
Introduction

Abstract Cultural number systems are the counting numbers *one*, *two*, *three*, their representations in various physical and spoken forms, and the arithmetic and mathematics performed with them. Different societies develop different number systems, and individual number systems change over time. Historically, this variability between and within cultures has been difficult to explain, especially when the human brain is considered to be the reason why number systems differ, the approach of the cognitive sciences. This book takes instead a novel archaeological approach: cultural number systems are studied here through their material component: the fingers, tallies, tokens, and written notations used to represent and manipulate numbers. By focusing on the material component, the archaeological approach is able to explain cross-cultural numerical variability as a matter of whether devices are used, which ones are used, and how they are used. The use of ethnographic data is also detailed. Early ethnographic sources provide important insight into what particular number systems were like before they were greatly influenced by Western contact, but such records must be used with caution.

Keywords Material engagement theory · Cognitive archaeology · Cultural number systems · Cross-cultural variability · Ethnographic data

Cultural number systems are the counting numbers *one*, *two*, *three*, their representations in various physical and spoken forms, and the arithmetic and mathematics performed with them, as developed by different societies in different times and places. This book is about cultural systems, and why they are so endlessly fascinating: their astonishing variability. Some societies count to very few numbers—perhaps no higher than *two* or *about three*—while others have unrestricted counting and might compute how high a number they could reach before the effort has exhausted all the energy in the universe and time itself.[1] Many societies organize their numbers by tens and twenties, amounts clearly traceable to counting on the

[1] Rotman (2000).

K. A. Overmann, *Cultural Number Systems*, Interdisciplinary Contributions to Archaeology, https://doi.org/10.1007/978-3-031-83383-0_1

fingers and toes, while others use anywhere from *two* to *seventy-four* as an organizing base. Some societies have writing, which turns out to be a highly concise material form for representing numbers; societies that don't write are often (quite wrongly) suspected of primitivism, but in some cases, it turns out their numbers are so complex they can't easily be represented by written forms.

I study cultural number systems as an archaeologist. While numbers are not the sort of thing that can be dug up during an excavation, being an archaeologist means I study numbers through a material lens. That is, I'm interested in the devices used to represent and manipulate numbers, as well as their role in conceptual emergence and elaboration. By using devices like fingers, tallies, tokens, and written notations, human societies have independently invented numbers many times over, and the cross-cultural results are both amazingly similar and impressively variable. As an example of something that is both similar and variable, consider finger-counting. Most societies count on their fingers—the only ones who don't are those who don't count as high as *five*—but no two societies count on their fingers in exactly the same way, even though everyone in any particular society counts on their fingers just like everyone else in that society does. Finger-counting and the other devices and behaviors used for numbers are important because they provide insight into how these conceptual systems develop, why they develop in only certain ways, and how local choices on what material forms to use and how to use them inject variability into the resultant content, organization, and structure.

As archaeologists go, I'm a *cognitive* archaeologist, someone who investigates the prehistory, function, and definition of the human mind. The theoretical framework I use, the Material Engagement Theory of Lambros Malafouris,[2] sees human thinking as involving material forms, and these influence *how* we think, including the concepts we form. Our ability to leverage material forms for cognitive purposes is arguably the key trait that defines us as a species and truly distinguishes us from other animals.[3] My approach to this work involves looking for patterns of change in the material record and then asking what those changes might imply about changes in behaviors and brains and what part the material forms may have played in them. In researching the material forms and behaviors used with numbers, I'm interested in what they can tell us about how we think, numerically speaking.

The inquiry is not without its challenges. For example, the further back in time we look, the less ancestral brains are like brains today, so their comparability becomes increasingly problematic. Conversely, the more recent in time we look, the less likely we are to find significant *evolutionary* change, the kind of genetically inheritable differences in brain form and function thought to occur over hundreds of thousands of years or longer. On the other hand—and this is one of the reasons I study number systems—the more recent in time we look, the more likely we are to detect *cultural* change in cognition. This is especially true for the last 5000 years, the period of demographic and technological intensification that began in the

[2] Malafouris (2013).

[3] Overmann (2021).

Neolithic and continues today. Not only are the behaviors and brains within this period effectively indistinguishable from our own, the material record is also much more robust, relative to remains that are 10,000 or more years old. Simply, the last 5000 years provide good insight into cultural change in cognition because there are fewer issues in cognitive comparability and more fine-grained archaeological data to provide insight into material, behavioral, and psychological change.

Only a few decades ago, archaeology was seen as having no role in understanding the ancient mind. Thoughts don't preserve, the thinking of the time held, so *what* ancient peoples once thought is "lost and gone for ever."[4] The best we might hope to do is infer ancient thinking from the material record. The problem with this is that sequential inferences about prehistoric belief systems become increasingly tenuous, no matter how well grounded they initially were in the material remains of the culture.[5] This is because beliefs are not isomorphic or reducible to the material objects that depict, realize, or anchor them, and the differences between the two can be significant and unpredictable.

But consider a system of physical weights. A system of increasingly heavy stones used as a standard for weighing implies a closely related conceptual system for weight (Fig. 1.1). This is because our lived experience of weight as a physical property, along with the devices we use to represent and manipulate it, inform our concepts of weight.[6] Objects weighing as much as the heaviest stone shown in the figure would be thought of as having that particular weight, and the amount would most likely be called by whatever name was given to the heaviest stone ("ounce," "gram," etc.). Because of the relation between the experience of weight, the devices used to represent and manipulate it, and the concepts formed about it, the system of stones used for weighing other objects provides some reasonably accurate insight into the associated concepts.

This is also true for numbers. Quantity, like weight, is another physical property, a condition of the planet we inhabit and the species that we are. As another physical

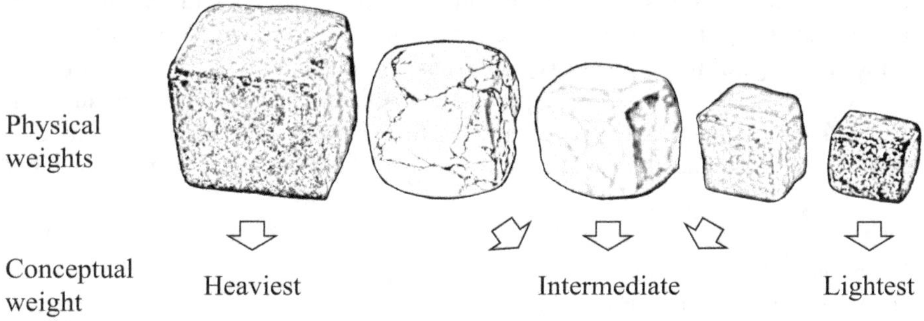

Fig. 1.1 Physical weights and associated concepts of weight. Image by the author

[4] Leach (1973), p. 768.

[5] Hawkes (1954).

[6] Renfrew (2001).

property, quantity has a similar potential for being represented and manipulated with material forms, and these forms have a similarly close relation with the concepts that are formed (a key difference being that quantity has a much greater potential for complex elaboration than weight does). Because they are physical properties, weight and quantity inform and unify cultural responses in ways that belief systems like cosmologies and religions do not. Thus, cultural systems like numbers are inherently comparable across cultures, no matter how many or significant the differences in other conceptual domains might be.[7]

Interestingly, not so long ago, psychology—like archaeology—was deemed not to have any role in understanding what numbers are as concepts and how societies get them. In the late nineteenth century, the mathematician Gottlob Frege (1848–1925) put this exclusion into blunt terms when he said that "psychology should not imagine it can contribute anything to the foundation of arithmetic."[8] This was because numbers have been seen as objective truths—"truths on a par with the truths of physics"[9]—and thus tractable to empirical discovery. This view of numbers as objective truths "out there" somewhere waiting to be discovered, "just like the fact that the sun has helium in it,"[10] can be traced back to ancient Greek philosophers like Plato.[11] As objective truths, numbers have been considered independent of the human mind. Like the sun's helium, they exist, and all we need to do is discover them and then try to understand them better. Since numbers were categorized as existing externally to and independently of the mind, and because mind and brain have been considered to be the same thing, it logically followed that brain-focused sciences like psychology had little to offer with respect to numbers—just as they have little role in explaining the gaseous makeup of the sun.

Nonetheless, not everyone agreed that this was a good explanation of what numbers are and how we get them. Cross-cultural variability is a good reason: if numbers are objective truths, then why do they differ according to the society that finds them, and why don't all societies find them? If numbers are not external to the mind in the way Plato said, then the mind—and hence, the brain—become the next logical place to look for answers.[12] A key question, then, becomes the brain's role in conceptualization and formatting. The psychologist Jean Piaget (1896–1980) took up the challenge of trying to answer this question. Piaget thought concepts of numbers were formed through the experience of quantity, with the brain imposing *invariant* logico-mathematical structure on that experience, regardless of its content.[13] This seems plausible, at least at first glance, because *two* of something added

[7] Overmann (2024a).

[8] Frege (1884), p. VI, as translated.

[9] Chomsky (2004), p. 43.

[10] Chomsky (2004), p. 43.

[11] Linnebo (2018); Maddy (1990).

[12] Brouwer (1981).

[13] Clarke and Beck (2021); Howard et al. (2018); Nieder (2016); Piaget (1928, 1950, 1952, 1970); also see critique in Overmann (2024b).

to *two* of something else make *four* things, irrespective of what those objects happen to be.

Nonetheless, there are several problems with Piaget's position.[14] One is that cognitive researchers have not conclusively defined or isolated the kind of innate logico-mathematical predispositions he predicted, though they have looked for them intently now for several decades. There is one exception, the ability to appreciate quantity that humans share with other species.[15] While it is generally known as "the number sense," this name is misleading, since we perceive *quantity*, rather than number; accordingly, I call it "the ability to perceive quantity" or "the perceptual system for quantity." Its function and neurological substrate have been well specified, though attempts to make it additionally responsible for computational processing have been unconvincing. Researchers have also largely failed to identify how and why human numerical thinking differs from that of other species. As a result, honeybees are said to have a concept of *zero*[16] and chimpanzees rational numbers.[17] Certainly, other species—mammals, birds, reptiles, amphibians, fish (all vertebrates) and perhaps even insects (invertebrates)—appear to share the ability to perceive quantity, as demonstrated and documented in a vast literature. Nonetheless, given the significant differences in the material and linguistic expressions of numbers between humans and other animals, only humans are likely to have concepts of numbers in the way we understand numbers to be. For example, no other species counts on its digits, which is where numbers start for humans.

Another problem with holding the brain responsible for numbers—especially when it is thought to impose *invariant* logico-mathematical structure—is that societies with very few numbers are placed at a serious cognitive disadvantage, relative to societies with highly elaborated numbers. Such explanations of the differences between societies may have sufficed in the nineteenth century, but they are woefully untenable in the twenty-first, especially in view of evidence demonstrating that the ability to appreciate quantity, thought to be the basis of numerical thinking, is one of the few human psychological capacities to work exactly the same for everyone across vast differences of culture.[18] That is, members of societies with almost no numbers perceive quantity in exactly the same way as Western people with highly elaborated numbers.[19]

In response to these and other problems, the ground is beginning to shift beneath the psychological paradigm that sees numbers as originating in the brain. The alternative to a brain-centric view is the one adopted here: the material devices used to represent and manipulate numbers—fingers, tallies, tokens, and written

[14] See Overmann (2024b).

[15] Dehaene (2011).

[16] Howard et al. (2018); Nieder (2016).

[17] Clarke and Beck (2021).

[18] Henrich et al. (2010).

[19] Gordon (2004).

notations—are the mechanism for their elaboration and the source of their various properties.[20] With this model, cross-cultural variability becomes a simple matter of *whether* material devices are used, *which* devices are used, and *how* those devices are used. Societies that don't use material devices to represent and manipulate numbers, as it turns out, have very few concepts of numbers. It's a difference in the conditions in which numerical concepts emerge and elaborate, not a difference in the ability to think.

Societies that do use material devices to represent and manipulate their numbers conceive of them in ways that correspond to the characteristics of the devices used. For example, counting on the fingers of one hand produces quinary (base-5) structure, counting on the fingers of both hands produces decimal (base-10) structure, counting on all the fingers and toes produces vigesimal (base-20) structure, and counting with the spaces between the fingers produces quaternary (base-4) or octal (base-8) structure, depending on how many hands are used. These conceptual outcomes are not differences in the ability to think; rather, they reflect different choices made in how to use the hands (and perhaps the feet) in counting. Understanding this means reversing the usual order of things: rather than seeing material forms as passive recipients of externalized mental content, we instead recognize our manuovisual engagement of material forms as a mechanism for generating ideas about numbers, like their structure.

A caveat is needed here: apart from this brief discussion in the introduction, this book isn't about the theories and methods I use in my numerical research. However, I will state some brief premises. First, my initial insight into the role of materiality in numerical cognition came from Lambros Malafouris, who observed that material forms make our ability to perceive quantity tangible and manipulable.[21] Second, I see the material forms used to represent and manipulate numbers as the source of their conceptual properties, and third, I treat the hand (when it is used in counting) and symbolic written numerals as material forms on par with physical devices like tallies and tokens.[22] The use of representational devices begins with the hand because of its visual salience, and what devices are capable of—and, importantly, what they are also incapable of—systematizes the recruitment of new devices for representing and manipulating numbers. To go beyond these premises, the interested reader can consult my other publications, where I spell these matters out in some detail.[23]

This book is instead a collection of short essays, each of which tries to explain why I find a particular cultural number system so very interesting. No principle guides the material collected here beyond the fact that the number systems appealed to me in some way. Sometimes it's because I see them as an archaeologist; by focusing on their material component, I can interpret them in a new way. My account of

---

[20] Overmann (2019, 2023).

[21] Malafouris (2010, 2013); also see discussion in Coolidge and Overmann (2012).

[22] Overmann (2016, 2019, 2023).

[23] E.g., Overmann (2023).

how the ancient kings of Persia once enumerated the men in their armies falls into this category (Chap. 6). Sometimes it's the thrill of having decoded a number system in some fashion, like figuring out how counting by sorting worked in Polynesia (Chap. 30). Often it's the astonishing beauty of a conceptual structure, the clever pragmatism of a counting method, or the shared humanity of numbers that I find so captivating. These analyses have taught me to see my own enculturated numbers, the Western tradition, as a predisposition for (wrongfully) thinking all numbers are like Western ones[24]—a tendency to be recognized and resisted.

Each essay gives a short account of what makes a particular number system stand out, and briefly explains the concepts and issues involved. I compare number systems across space and time where I think they will be useful in illustrating a particular point; these comparisons also serve to emphasize the commonality of our numerical adaptations. The cultural systems are organized into six geographical areas (Near East, Africa, Europe, Asia, Oceania, and the Americas) by considerations of time (oldest before newest) and space (e.g., north to south). The result is an eclectic, non-exhaustive look at cultural number systems as they are found around the planet and throughout time.

It's important to address the use of ethnographic material, as it is a primary resource for my analyses. Early sources are particularly valuable, as they often provide the only insight into a cultural number system before it was greatly influenced by Western contact. Here too the interested reader might consult some of my earlier publications where the issues are spelled out in detail.[25] But in brief, the use of such data is often questioned. The first issue is one of legitimacy. The issue has to do with making comparisons between archaic/prehistoric and modern societies; seeing a modern society as some kind of "latter-day survival" of an archaic/prehistoric one has been called "immoral."[26] While is certainly true that, in general, human behaviors and belief systems are incredibly variable,[27] there are things we do all share, like the need for food and shelter and the tendency to organize ourselves cooperatively in getting them. This sharedness is certainly true of numbers. Societies do very similar things to represent and manipulate their numbers, despite being widely separated by space, time, circumstance, culture, and language.[28]

Another worry is the reliability of the ethnographic data, especially the early material. This concern recognizes the vast gulf between present-day scientific practices and ethical norms and their relative (or complete) absence, say, a century ago.[29] Early ethnographic accounts were written by an eclectic mix of people: "explorers, fur trappers, traders, political revolutionaries, escaped convicts, a nobleman or two, several knights, military men, schoolteachers, a few mathematicians, missionaries,

[24] Rotman (2000).

[25] E.g., Overmann (2023, 2024a).

[26] Gosden (1999), p. 9.

[27] Smith (1955).

[28] Divale (1999).

[29] Currie (2016); Ember and Ember (2000); Kaplan and Manners (1972); Pelto and Pelto (1978).

clergymen, priests, a bishop later canonized as a saint, and a humanitarian who would win the Nobel Peace Prize."[30] Collectively, they lacked training, experience, and the mandate and motivation to collect accurate data. Their writing, moreover, is often biased by notions of Western cultural superiority, and this bias influenced both what they chose to record and the language they used to record it. Nonetheless, when the observations are considered cumulatively, they provide their own solution to the problem at hand: "when even the most problematic of instruments find the same thing, time and time again, then what they report seems likely to be a case of what is actually there."[31]

There is also an assumption—one that seems widely held—that because early observers were eclectic and untrained, their reports were always and incurably unreliable and should be immediately discarded as such. Yet when any particular observer is investigated, it often turns out that he (it's usually a man) had significant experience with an Indigenous society and regarded its people quite favorably. An example is Christian Schrumpf (1818–1884), the nineteenth-century missionary who described an African method of counting in which different Basuto men kept track of the units, tens, and hundreds (Chap. 12).[32] When Schrumpf wrote about this method, he had 16 years of experience among the Basuto,[33] and as a missionary, he likely had close, daily contact with multiple individuals in both public and private spaces. Granted, he found the method primitive, describing it biased language.[34] But I have chosen to ignore the bias, and reinterpret the observation accordingly. Where Schrumpf saw a difficulty in counting, I see men collaborating to form a living abacus.[35] This complex, cooperative type of finger-counting is important, for it can explain how concepts like exponents (groups of tens, hundreds, thousands, etc.) are possible in the seeming absence of formal devices (e.g., physical abacuses; counting boards). That is, accounts like the one from Schrumpf reveal numerical concepts as having a material basis, and not a purely mental or linguistic one.

Though I draw upon "early" ethnographies (defined here as published before 1961, the average year of publication for the 719 references cited in the book), I don't take them at face value. Instead, I check their contents against more recent publications, which make up the bulk of the references cited in the book. Of the 719 references, 518 (72.0%) were published in 1961 or thereafter, with 161 (22.4%) published within the last ten years. This same proportionality is borne out in the different chapters. For example, the chapter on Australian number systems (Chap. 28) cites 23 references, of which 17 (73.9%) were published after 1961 and 7 (30.4%) were published in the past decade. One of the early ethnographies cited in

30 Overmann (2019), p. 68.

31 Overmann (2024a), p. 422.

32 Schrumpf (1862).

33 Jousse (1889).

34 Schrumpf (1862).

35 Overmann (2020).

the chapter was published in 1852;[36] to confirm that it accurately portrayed Indigenous life, I consulted three sources from different periods (1886, 1966, and 2017).[37] Similarly, I checked the information in the other early sources in the chapter (Dawson, 1881; Howitt, 1904; Mathews, 1904; Tindale, 1925) against more recent publications (e.g., Blake et al., 1998; Bowern & Zentz, 2012; Harris, 1982, 1987; Mayer, 2018; Zhou & Bowern, 2015). I hope this insight shows my approach to this research to be careful, thorough, and ethical.

Beyond the issue of reliability is the question of whether Indigenous peoples have the right to control the information published about them. If these accounts were contemporary stories, the answer would be an unequivocal "yes, of course they do." For historical data, the situation is less clearcut, as the early material does not always clearly align with specific cultural groups as they define themselves today. While I have tried to empower Indigenous voices and cultural systems throughout, they generally appear as observed and recorded by early travelers and missionaries, whose mere presence on the cultural landscape was already acting to disturb it. No matter how well scrubbed this material may now be of biases, the very fact of its inclusion may be culturally insensitive. I sincerely hope not, and I also hope that my approach and admiration will transcend any sensitivities I have inadvertently touched.

A final issue has to do with the selection of ethnographic material. As I noted earlier, my guiding principle was personal interest. To be included, a number system had to stand out in some way against the range of cultural systems that I know, and I know more than a few. The references also had to withstand scrutiny. Despite this care, there are omissions in my account. They are related either to my lack of knowledge—my familiarity with cultural number systems is better than average, but neither omniscient nor infallible—or to my seeing, for example, prototypical decimal systems as bland and unremarkable. Every reader who knows an interesting number system not mentioned somewhere in the book is hereby invited to get in touch immediately with all the details; I promise to be immensely interested.

When I first started this research, I thought about numbers the way I suppose most people do, in terms of the Western ones I learned in school.[38] Yet I have come to see Western numbers as just one cultural system among many, albeit an old and highly elaborated cultural system. If Western numbers have become the most highly elaborated numerical tradition that humans have yet developed,[39] as numbers go, they are nonetheless narrow and bland, compared to the range of how numbers are conceived cross-culturally. Western numbers increase by adding *one* to regular multiples of *ten*, an additive procedure formalized as the successor function that sends one natural number to the next. In other cultural systems, numbers anticipate, backtrack, and concatenate in ways that require users to keep mental track of complex

[36] Morgan (1852).

[37] Curr (1886); Flannery (2017); Tipping (1966).

[38] Rotman (2000).

[39] Gowers (2008).

multiplications, subtractions, and sometimes even divisions; the levels of complexity are such that people enculturated into the Western tradition, with its very simple numbers, would find quite difficult to do. If there is an overarching message to this book—beyond the simple fact that the cross-cultural variability found in cultural number systems is interesting—it's that the human capacity for numbers cannot be fully appreciated unless we consider its cross-cultural variability.

While I hope readers will enjoy the material collected here on its own merits, I also hope they will come away from reading the book with the understanding that the human capacity for numbers is so much wider and more complicated than the narrow and bland Western decimal tradition makes it appear. This wide and shared capacity for numbers is something we ought to celebrate and explore.

References

Blake, B. J., Clark, I. D., & Krishna-Pillay, S. H. (1998). Wathawurrung and the Colac language of Southern Victoria. In B. J. Blake (Ed.), *Wathawurrung: The language of the Geelong–Ballarat area* (pp. 59–154). Pacific Linguistics.

Bowern, C., & Zentz, J. (2012). Diversity in the numeral systems of Australian languages. *Anthropological Linguistics, 54*(2), 133–160.

Brouwer, L. E. J. (1981). *Brouwer's Cambridge lectures on intuitionism* (D. van Dalen, Ed.). Cambridge University Press.

Chomsky, N. (2004). *The generative enterprise revisited: Discussions with Riny Huybregts, Henk van Riemsdijk, Naoki Fukui and Mihoko Zushi* (2nd ed.). Mouton de Gruyter.

Clarke, S., & Beck, J. (2021). The number sense represents (rational) numbers. *Behavioral and Brain Sciences, 44*, 1–57.

Coolidge, F. L., & Overmann, K. A. (2012). Numerosity, abstraction, and the emergence of symbolic thinking. *Current Anthropology, 53*(2), 204–225.

Curr, E. M. (1886). *The Australian race: Its origin, languages, customs, place of landing in Australia, and the routes by which it spread itself over that continent* (Vol. 2). John Ferres, Government Printer.

Currie, A. (2016). Ethnographic analogy, the comparative method, and archaeological special pleading. *Studies in History and Philosophy of Science, Part A, 55*, 84–94.

Dawson, J. (1881). *Australian Aborigines: The languages and customs of several tribes of Aborigines in the western district of Victoria, Australia*. George Robertson.

Dehaene, S. (2011). *The number sense: How the mind creates mathematics* (Revised). Oxford University Press.

Divale, W. (1999). Climatic instability, food storage, and the development of numerical counting: A cross-cultural study. *Cross-Cultural Research, 33*(4), 341–368.

Ember, M., & Ember, C. R. (2000). Testing theory and why the "units of analysis" problem is not a problem. *Ethnology, 39*, 349–363.

Flannery, T. (2017). Introduction. In T. Flannery (Ed.), *The life and adventures of William Buckley: Text classics* (pp. 1–12). Text Publishing.

Frege, G. (1884). *Die Grundlagen der Arithmetik. Eine logisch mathematische Untersuchung über den Begriff der Zahl [The foundations of arithmetic. A logical-mathematical investigation into the concept of number]*. Wilhelm Koebner.

Gordon, P. (2004). Numerical cognition without words: Evidence from Amazonia. *Science, 306*(5695), 496–499.

Gosden, C. (1999). *Anthropology & archaeology: A changing relationship*. Routledge.

Gowers, T. (Ed.). (2008). *The Princeton companion to mathematics*. Princeton University Press.

Harris, J. (1982). Facts and fallacies of Aboriginal number systems. *Works Papers of SIL-AAB, Series B, 8*, 153–181.

Harris, J. (1987). Australian Aboriginal and Islander mathematics. *Australian Aboriginal Studies, 2*, 29–37.

Hawkes, C. (1954). Archeological theory and method: Some suggestions from the Old World. *American Anthropologist, 56*(2), 155–168.

Henrich, J., Heine, S. J., & Norenzayan, A. (2010). The weirdest people in the world? *Behavioral and Brain Sciences, 33*(2–3), 61–135.

Howard, S. R., Avarguès-Weber, A., Garcia, J. E., Greentree, A. D., & Dyer, A. G. (2018). Numerical ordering of zero in honey bees. *Science, 360*(6393), 1124–1126.

Howitt, A. W. (1904). *The native tribes of south-east Australia*. Macmillan and Co.

Jousse, T. (1889). *La mission française évangélique au sud de l'Afrique son origine et son développement jusqu'a nos jours [The French evangelical mission in South Africa: Its origin and development to the present day]* (Vol. 2). Librairie Fischbacher.

Kaplan, D., & Manners, R. A. (1972). *Culture theory*. Prentice-Hall.

Leach, E. R. (1973). Concluding address. In C. Renfrew (Ed.), *The explanation of culture change: Models in prehistory. Proceedings of a meeting of the Research Seminar in Archaeology and Related Subjects held at the University of Sheffield, December 14–16, 1971* (pp. 761–771). Gerald Duckworth.

Linnebo, Ø. (2018). Platonism in the philosophy of mathematics. In E. N. Zalta (Ed.), *Stanford Encyclopedia of Philosophy (Spring 2021)*. Stanford University. https://plato.stanford.edu/archives/spr2018/entries/platonism-mathematics/

Maddy, P. (1990). *Realism in mathematics*. Clarendon Press.

Malafouris, L. (2010). Grasping the concept of number: How did the sapient mind move beyond approximation? In C. Renfrew & I. Morley (Eds.), *The archaeology of measurement: Comprehending heaven, earth and time in ancient societies* (pp. 35–42). Cambridge University Press.

Malafouris, L. (2013). *How things shape the mind: A theory of material engagement*. MIT Press.

Mathews, R. H. (1904). Language of the Wuddyāwūrru Tribe, Victoria. *Zeitschrift für Ethnologie, 36*(6), 729–734.

Mayer, C. J. (2018). *Numeral systems in New Guinea*. https://www.researchgate.net/publication/322506379_Numeral_Systems_in_New_Guinea

Morgan, J. (1852). *The life and adventures of William Buckley: Thirty-two years a wanderer amongst the Aborigines of the then unexplored country around Port Phillip, now the Province of Victoria* (19th ed.). Griffin Press.

Nieder, A. (2016). Representing something out of nothing: The dawning of zero. *Trends in Cognitive Sciences, 20*(11), 830–842.

Overmann, K. A. (2016). The role of materiality in numerical cognition. *Quaternary International, 405*, 42–51.

Overmann, K. A. (2019). *The material origin of numbers: Insights from the archaeology of the ancient Near East*. Gorgias Studies in the Ancient Near East 14. Gorgias Press.

Overmann, K. A. (2020). The curious idea that Māori once counted by elevens, and the insights it still holds for cross-cultural numerical research. *Journal of the Polynesian Society, 129*(1), 59–84.

Overmann, K. A. (2021). The material difference in human cognition. *Adaptive Behavior, 29*(2), 123–136.

Overmann, K. A. (2023). *The materiality of numbers: Emergence and elaboration from prehistory to present*. Cambridge University Press.

Overmann, K. A. (2024a). Prehistoric numeracy: Approaches, assumptions, and issues. In T. Wynn, K. A. Overmann, & F. L. Coolidge (Eds.), *The Oxford handbook of cognitive archaeology* (pp. 411–432). Oxford University Press.

Overmann, K. A. (2024b). Strange and wonderful: Numbers through a new (material) lens. *Cuneiform Digital Library Journal, 2*, 1–21.

Pelto, P. J., & Pelto, G. H. (1978). *Anthropological research: The structure of inquiry* (2nd ed.). Cambridge University Press.

Piaget, J. (1928). Logique génétique et sociologie [Genetic logic and sociology]. *Revue Philosophique de la France et de l'Étranger, 105*, 167–205.

Piaget, J. (1950). *Introduction à l'epistemologie genetique: La pensée mathématique [Introduction to genetic epistemology: Mathematical thinking]*. Presses Universitaires de France.

Piaget, J. (1952). *The child's conception of number* (C. Gattegno & F. M. Hodgson, Trans.). Routledge.

Piaget, J. (1970). *Genetic epistemology* (E. Duckworth, Trans.). W. W. Norton & Company.

Renfrew, C. (2001). Commodification and institution in group-oriented and individualizing societies. *Proceedings of the British Academy, 110*, 93–117.

Rotman, B. (2000). *Mathematics as sign: Writing, imagining, counting*. Stanford University Press.

Schrumpf, C. (1862). Sessuto. Ein Beitrag zur Süd-Afrikanischen Sprachenkunde, geschöpft aus meiner sechszehnjährigen Erfahrung [Sessuto. A contribution to South African linguistics, drawn from my sixteen years of experience]. *Zeitschrift der Deutschen Morgenländischen Gesellschaft, 16*(3), 448–481.

Smith, M. A. (1955). The limitations of inference in archaeology. *The Archaeological News Letter, 6*, 3–7.

Tindale, N. B. (1925). Natives of Groote Eylandt and of the west coast of the Gulf of Carpentaria, Part 2. *Records of the South Australian Museum, 3*(2), 103–134.

Tipping, M. J. (1966). Buckley, William (1780–1856). *Australian Dictionary of Biography*. https://adb.anu.edu.au/biography/buckley-william-1844/text2133

Zhou, K., & Bowern, C. (2015). Quantifying uncertainty in the phylogenetics of Australian numeral systems. *Proceedings of the Royal Society B: Biological Sciences, 282*(1815), 20151278.

Part I
Number Systems in the Ancient Near East

Chapter 2
Sumerian Numbers

Abstract The world's earliest unambiguous numbers come from Mesopotamia, the ancient civilization that once flourished between the Tigris and Euphrates Rivers. These numbers are usually assigned to the Sumerians, an early cultural group. The Sumerian number system was sexagesimal, organized on the number *sixty*. While we may never know why the Sumerians chose *sixty* instead *ten*, they counted on their fingers in the typical manner: five on each hand, as revealed by their words for numbers. These words were not recorded until long after writing had been invented because numerical notations work better when they aren't cluttered with phonetic clues. The Sumerians also counted different kinds of objects differently—sheep were counted one way, food rations in another, containers of grain in a third way. Such object-specified counting is not uncommon, as it is also found in places like Polynesia. Unlike Polynesia, where object-specified counting involved the goods being enumerated, the Sumerians used tokens and numerical notations to represent their numbers, and these technologies enabled the elaboration of numbers into one of the ancient world's greatest mathematical traditions.

Keywords Sexagesimal · Object-specified counting · Polyvalence · Phonography · Tokens

The world's earliest unambiguous numbers are found in Mesopotamia, the ancient civilization that flourished in the fertile crescent between the Tigris and Euphrates Rivers during the Neolithic and Bronze Age. Though it's difficult to determine exactly who invented these numbers, they are most often associated with one of the region's earliest cultural groups, the Sumerians. This number system is known for several odd traits, like using *sixty* as its base and counting different types of objects differently. If these traits become less mysterious when they are compared to contemporary number systems, four new surprises about Sumerian numbers also emerge.

The Sumerians counted on their fingers just like we do: five on each hand. This is surprising because Sumerian numbers were sexagesimal, or base-60, which is

K. A. Overmann, *Cultural Number Systems*, Interdisciplinary Contributions to Archaeology, https://doi.org/10.1007/978-3-031-83383-0_2

rare among the world's number systems.[1] We know they counted on their fingers like we do because Sumerian spoken numbers up to *sixty* were formed with fives, tens, and twenties, numbers that emerge from the use of one or both hands and the toes in counting.[2] In Sumerian, the word for *nine* meant *five-and-four*, which is all the fingers on one hand plus four on the other; *thirty* was *three-tens*, the number of three hands; *forty* was *two-twenties*, the total digits of two men; and *fifty* was *forty-and-ten*, all the fingers and toes of two men plus the ten fingers of a third.[3]

Why the Sumerians used 6 and not 10 as the next higher base remains a mystery. Possibly it was motivated by the increase in factorability, as 60 can be evenly divided by 3 and 6, which isn't true of 100, the next-higher decimal base. Every now and then, someone suggests the Sumerians reached 60 by counting the three segments on each of the four fingers to realize 12 and then multiplying the result by 5.[4] However, counting segments instead of fingers produces *duodecimal* or base-12 systems that are organized quite differently than the Sumerian numbers were. For example, in the Nigerian language Birom, the term for *nine* means *twelve-less-three*, *twenty* is *twelve-and-eight*, and *thirty* is *two-twelves-and-six*.[5] The important point is that in a duodecimal system, *sixty* is not productive (that is, used as the basis of higher numbers), so in Iten (or Aten), another Nigerian language with duodecimal numbers, *sixty* is *five-twelves*, *seventy* is *five-twelves-and-ten*, *eighty* is *six-twelves-and-eight*, *ninety* is *seven-twelves-and-six* and *one hundred* is *eight-twelves-and-four*.[6]

No instance of using the fingers to count to 12 and then using five cycles of 12 as a basis for sexagesimal organization is known among the world's number systems. As noted earlier, sexagesimal organization is quite rare. Only one other possible instance is known, the Ekari (Kapauku) system in New Guinea (Chap. 27), and its basis appears to be counting on the body—a kind of counting that involves body parts other than (and in addition to) the fingers and toes. Both the Sumerian and Ekari number systems are recent enough that their number-words don't just have unambiguous etymological roots in counting with the fingers and other body parts, they also provide good insight into the way these things were used in counting. Just to be clear: in neither case was it counting five cycles of 12 with the finger segments.

The Sumerian words for numbers weren't recorded until long after writing was available. This is surprising because writing began with numbers in the form of "tokens," simple geometric shapes made of clay (Fig. 2.1). Around the middle of the fourth millennium BCE, tokens began to be enclosed in clay spheres known as *bullae*.[7] Often the exterior surfaces of bullae were impressed with the tokens they

[1] Comrie (2013).
[2] Epps (2006).
[3] Blažek (1999), Edzard (1980).
[4] E.g., Hanson (2015); Huylebrouck (1997); Petrov (2023); Thomas (1920).
[5] Bouquiaux (1962).
[6] Bouquiaux (1964).
[7] Amiet (1966).

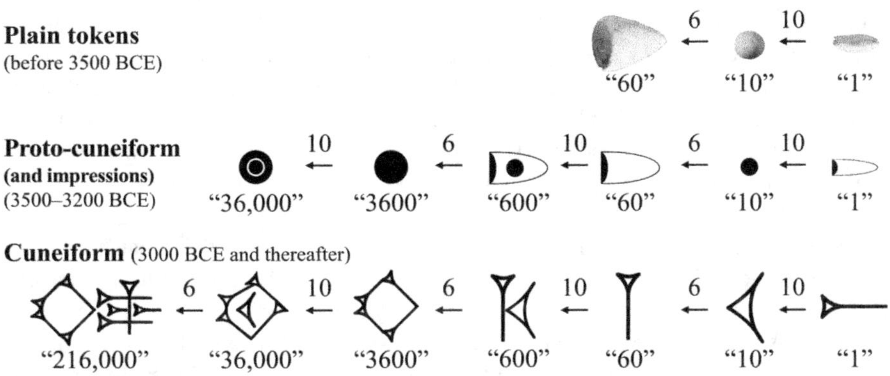

Fig. 2.1 Technologies for representing numbers in Mesopotamia. While only one counting sequence is shown to aid comparability, more than a dozen were used. Tokens adapted from Englund (2006, Fig. 24, p. 29); proto-cuneiform notations redrawn from Nissen et al. (1993, Fig. 28, p. 28). A version was previously published in Overmann (2023, Fig. 13.2, p. 313)

enclosed, a practice that avoided the need to break them open to recover their contents. Within just a few centuries, token-shaped impressions were being marked with a stylus on flat clay tablets and labeled with small pictures, becoming a true system of (hand)writing.[8]

While numbers were foundational to these developments, Sumerian words for small numbers weren't recorded for centuries, and words for large numbers didn't appear for more than a thousand years.[9] After writing emerged, techniques for recording sound values—*phonography*—had to be invented, as this differs from using small pictures that mean what they look like. Once phonography was available, there also had to be a reason to apply it to numbers, signs that simply work better *as* numbers without the complexity that sound values add.[10] We can see this same phenomenon for ourselves by comparing phonetic numbers with their non-phonetic counterparts: διακόσια τριάντα τέσσερα (Greek) and 234. Numerical signs without phonetic encoding are more concise, which makes them easier to understand and manipulate and also improves their capacity for revealing relations and patterns. In Mesopotamia, sound values were eventually recorded by foreign scribes learning the Sumerian words for numbers (*diš*, *ia*, *u*, *ninnu*, *g̃eš*) in addition to the Sumerian numerals (𒁹, 𒐊, 𒌋, 𒐐, 𒁹). A modern counterpart would be our learning the Latin words for numbers (*unus*, *quinque*, *decem*, *quinquaginta*, *sexaginta*) in addition to the Roman numerals (I, V, X, L, LX).

Sumerian numbers counted different types of items differently, but this isn't unusual. Sumerians counted sheep one way, food rations another, and containers of

[8] Schmandt-Besserat (1992).

[9] Damerow (1988).

[10] Overmann (2022).

grains like barley a third.[11] This "object-specified counting" has been interpreted as indicating a rudimentary concept of number, essentially the claim that Sumerians didn't really comprehend numbers or had only a fragmented understanding of them.[12] Yet surprisingly, this kind of counting is actually fairly common around the world, as examples are found in Africa, Oceania, and the Americas. Such counting is also associated with highly elaborated number systems, the very opposite of being rudimentary.[13]

For example, object-specified counting and the counting method that it emerged from are found in Polynesia (also see Chap. 30). In this Polynesian "tally-counting," every tenth item is set aside to represent *ten* of the items being counted. The unit of counting also varies, depending on the object's shape and size, with objects counted singly, by pairs, or in groups of *four*. This counting-by-sorting with multiple countable units creates object-specified counting sequences. When different Polynesian counting sequences are considered together, including a unique binary variant from the island of Mangareva,[14] they look a lot like what we find in Mesopotamia. In fact, if we were to replace the Polynesian words with token shapes and sizes (Fig. 2.2), the result greatly resembles the early Sumerian counting sequences.[15] Similarities include characteristic upshifts in value, where the numerical values of a sequence of signs are multiplied by a common factor like *two*; *polyvalence*, where the same sign would have different numerical values according to what it was counting; and the use of "round numbers," ignoring the remainders of all but the final step of sorting.[16] These similarities suggest that the Sumerians likely used a counting method similar to the one found in Polynesia.

Such counting practices are motivated by two things, conditions that were also true in ancient Sumer. First, items that are socially important need to be counted frequently and in large quantities.[17] Second, counting needs to be easy, accurate, and involve minimal amounts of physical and mental effort.[18] Counting by sorting does both extremely well, especially when things are counted in pairs or by fours. Today we tend not to think about such practical considerations, since we either don't manually count large quantities of physical objects very often, or we use automated methods when we do.

Sumerian numbers were represented by tokens and numerical notations. This in itself isn't surprising, since these material forms are what give us insight into the ancient numbers in the first place. What is surprising, however, is how much we tend to underestimate their importance. In Polynesia, once counting-by-sorting was

---

[11] Nissen et al. (1993).

[12] Damerow (1996); Schmandt-Besserat (1992).

[13] Overmann (2021).

[14] Bender and Beller (2014).

[15] Overmann (2021).

[16] De Odorico (1995).

[17] Bender et al. (2015).

[18] Overmann (2020).

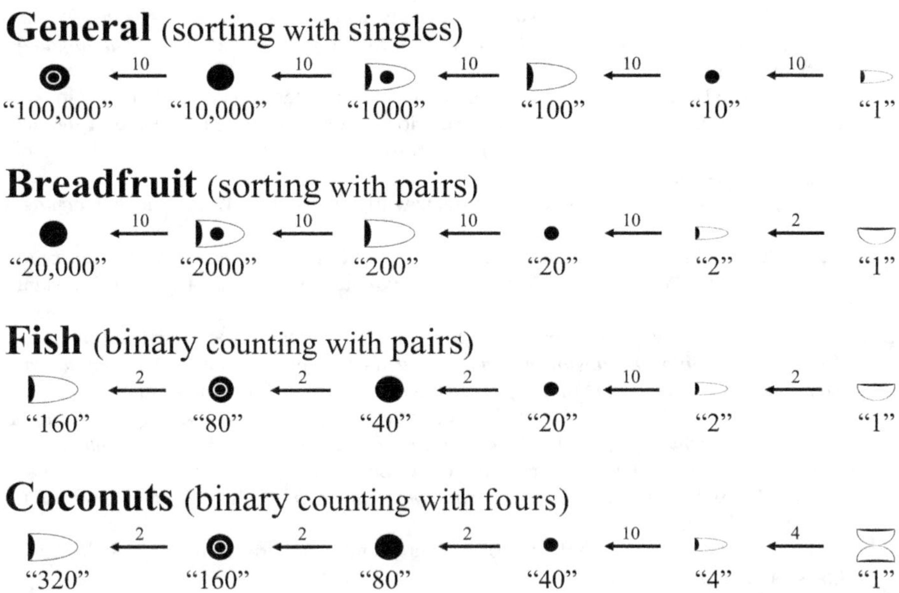

Fig. 2.2 Mangarevan words in object-specified counting replaced by hypothetical tokens; the result resembles the early Sumerian counting sequences. Adapted from Overmann (2021, Fig. 4b, p. 307)

completed, the amounts were kept track of with words and human memory. In comparison, material objects like notations and tokens could be used away from the objects being counted. This facilitated recording and calculating, as well as the gradual refinement and complexity of both.[19] Material objects like tokens and notations are what enabled Sumerian numbers to be developed into one of the ancient world's greatest mathematical traditions.

References

Amiet, P. (1966). Il y a 5000 ans les Élamites inventaient l'écriture [5000 years ago the Elamites invented writing]. *Archéologia, 12*, 16–23.

Bender, A., & Beller, S. (2014). Mangarevan invention of binary steps for easier calculation. *Proceedings of the National Academy of Sciences of the United States of America, 111*(4), 1322–1327.

Bender, A., Schlimm, D., & Beller, S. (2015). The cognitive advantages of counting specifically: A representational analysis of verbal numeration systems in Oceanic languages. *Topics in Cognitive Science, 7*(4), 552–569.

[19] Overmann (2021).

Blažek, V. (1999). *Numerals: Comparative etymological analyses and their implications: Saharan, Nubian, Egyptian, Berber, Kartvelian, Uralic, Altaic and Indo-European languages*. Masarykova Univerzita.

Bouquiaux, L. (1962). A propos de numération: L'emploi du système décimal et du système duodécimal dans la langue birom (Nigéria septentrional) [On numeration: The use of the decimal system and the duodecimal system in the Birom language (northern Nigeria)]. *Africana Linguistica, 1*, 7–10.

Bouquiaux, L. (1964). A word list of Aten (Ganawuri). *Journal of West African Languages, 12*, 5–25.

Comrie, B. (2013). Numeral bases. In M. S. Dryer & M. Haspelmath (Eds.), *World atlas of language structures online*. Max Planck Institute for Evolutionary Anthropology. http://wals.info/chapter/131

Damerow, P. (1988). Individual development and cultural evolution of arithmetical thinking. In S. Strauss (Ed.), *Ontogeny, phylogeny, and historical development: The Tel Aviv annual workshop in human development* (Vol. 2, pp. 125–152). Ablex Publishing Corporation.

Damerow, P. (1996). Prehistory and cognitive development. In M. Killen (Ed.), *Invited lecture at the twenty-fifth annual symposium of the Jean Piaget Society Berkeley, June 1 – June 3, 1995* (pp. 1–37). Max Planck Institute for the History of Science.

De Odorico, M. (1995). *The use of numbers and quantifications in the Assyrian royal inscriptions*. University of Helsinki.

Edzard, D. O. (1980). Sumerisch 1 bis 10 in Ebla [Sumerian 1 to 10 in Ebla]. *Studi Eblaiti, III*(5–8), 121–127.

Englund, R. K. (2006). An examination of the "textual" witnesses to the Late Uruk world systems. In Y. Gong & Y. Chen (Eds.), *Special issue of oriental studies: A collection of papers on ancient civilizations of Western Asia, Asia Minor and North Africa* (pp. 1–38). Oriental Studies.

Epps, P. (2006). Growing a numeral system: The historical development of numerals in an Amazonian language family. *Diachronica, 23*(2), 259–288.

Hanson, R. (2015). The origins of the number 60 as a counting method. *Stuff*. https://www.stuff.co.nz/science/71273903/roger-hanson-the-origins-of-the-number-60-as-a-counting-method

Huylebrouck, D. (1997). Tellen op de handen in Afrika en de oorsprong van het duodecimale system [Counting on hands in Africa and the origins of the duodecimal system]. *Wiskunde En Onderwijs, 89*, 53–57.

Nissen, H. J., Damerow, P., & Englund, R. K. (1993). *Archaic bookkeeping: Early writing and techniques of economic administration in the ancient Near East* (P. Larsen, Trans.). University of Chicago Press.

Overmann, K. A. (2020). The curious idea that Māori once counted by elevens, and the insights it still holds for cross-cultural numerical research. *Journal of the Polynesian Society, 129*(1), 59–84.

Overmann, K. A. (2021). A new look at old numbers, and what it reveals about numeration. *Journal of Near Eastern Studies, 80*(2), 291–321.

Overmann, K. A. (2022). Early writing: A cognitive archaeological perspective on literacy and numeracy. *Visible Language, 56*(1), 8–44.

Overmann, K. A. (2023). *The materiality of numbers: Emergence and elaboration from prehistory to present*. Cambridge University Press.

Petrov, P. (2023). Systems of cardinal numerals in languages around the world. *Linguistics & Polyglot Studies, 9*(3), 82–105.

Schmandt-Besserat, D. (1992). *Before writing: From counting to cuneiform*. University of Texas Press.

Thomas, N. W. (1920). Duodecimal base of numeration. *Man, 20*, 25–29.

Chapter 3
Minority Traditions

Abstract The textual record of Mesopotamia hints at two minority number systems, or number systems associated with relatively small, non-dominant cultural groups. One of these possible minority systems counted by threes in a way that is comparable to contemporary counting systems from Papua New Guinea (Waskia and Bukiyip). Ternal numbers are found on some tablets from the Early Old Babylonian and Neo-Babylonian periods. A second possible minority number system may have been an *eme-sal*, a dialect of Sumerian spoken by women. Today, such gendered differences in language are found all over the world, with the possible exception of Africa. Eme-sal numbers are found in some Sumerian literary texts, where they were used to characterize certain speech as female in origin. Today, minority number systems are in danger of disappearing, for reasons that include replacement by number systems from dominant cultural groups or adoption as a way of maintaining parity in politics and trade. However, in these cases, the words for the smallest numbers—*one*, *two*, and *three*—are usually retained for reasons having to do with perception and memory.

Keywords Ternal counting · Counting by threes · Eme-sal · Gendered speech · Minority traditions

The ancient Near East may have included a couple of minority numerical traditions, in addition to the more well-established (and today, better-known) number systems of the Sumerians, Akkadians, and Elamites. Such minority traditions have a couple of interesting characteristics. First, they may count to fairly low numbers, so when they come into contact with systems that count to higher numbers, they tend to borrow those higher numbers. Second, they are often associated with smaller, less dominant cultural groups. This means they might adopt a foreign number system, either to ensure equity and fairness in trade or to acquire prestige from using the numbers of a more powerful cultural group. Foreign numbers might also be imposed

K. A. Overmann, *Cultural Number Systems*, Interdisciplinary Contributions to Archaeology, https://doi.org/10.1007/978-3-031-83383-0_3

Table 3.1 Ternal counting system (Balke, 2010, Table 1, p. 46)

Nr.	Ternal number	Meaning	Nr.	Ternal number	Meaning
1	*bi*	1	7	*peš-peš-bi*	3 + 3 + 1
2	*bi-bi*	1 + 1	8	*peš-peš-bi-bi*	3 + 3 + 1 + 1
3	*peš*	3	9	*peš-peš-peš*	3 + 3 + 3
4	*peš-bi*	3 + 1	10	*peš-peš-peš-bi*	3 + 3 + 3 + 1
5	*peš-bi-bi*	3 + 1 + 1	11	*peš-peš-peš-bi-bi*	3 + 3 + 3 + 1 + 1
6	*peš-peš*	3 + 3	12	*peš-peš-peš-peš*	3 + 3 + 3 + 3

through conquest. Minority numbers can and often do disappear under these conditions.[1]

Minority numerical traditions aren't surprising. Mesopotamia was a large geographic region that was both culturally and ethnically diverse, and well connected both internally and externally through things like commerce and warfare. Any geographic region with these characteristics has the potential to contain a lot of different numerical traditions. It's both possible and likely that we would find only the more dominant numerical traditions in the archaeological and textual records because very little survives over thousands of years, so only the most prevalent items tend to be found after thousands of years.

One minority tradition counted by threes. This "ternal" counting system (Table 3.1) appears occasionally in lexical, literary, and ritual texts, things that scribes or priests are most likely to have written.[2]

Tablets with ternal numbers are found from the beginning of the Early Old Babylonian period (2000 BCE) to the end of the Neo-Babylonian period (539 BCE). Though these tablets come from cities from both Sumer and Akkad, scholars have most often associated them with the Sumerian culture.[3] Some have speculated that ternal numbers represented an early, archaic form of the Sumerian numbers.[4] However, this is unlikely, for two reasons. First, the spoken forms of the smallest numbers in any number system (*one* through *five*, or in this case, *bi*, *bi-bi*, *peš*, *peš-bi*, *peš-bi-bi*) tend to preserve over long periods of time. This means that if the ternal numbers were related to the Sumerian language, they should be similar to its smallest numbers (*diš*, *min*, *eš*, *limmu*, *ia*)—but they aren't. Second, ternal numbers are known only in phonetic form, never as tally marks or tokens. This suggests that ancient writers knew them primarily through speech, rather than record-keeping or calculating. Other scholars have proposed that ternal counting "was reserved for strictly household usage and employed mostly by women."[5] Such usage wouldn't exclude the possibility of a minority tradition.

---

[1] Eberhard et al. (2024).

[2] Balkc (2010); Civil (2005); Lambert (1969).

[3] Balke (2010); Civil (2005); Lambert (1969).

[4] Thureau-Dangin (1928).

[5] Diakonoff (1983), p. 90.

Table 3.2 Waskia pair-counting system (Ross & Paol, 1978, p. 60)

Nr.	Waskia number	Gloss	Nr.	Waskia number	Gloss
1	*itoketa*	1	5	*itelala-itelala-itoketa*	2 + 2 + 1
2	*itelala*	2	6	*iteltoke-iteltoke*	3 + 3
3	*iteltoke*	2 + 1	7	*iteltoke-iteltoke-itoketa*	3 + 3 + 1
4	*itelala-itelala*	2 + 2	8	*iteltoke-iteltoke-itelala*	3 + 3 + 2

Similar counting systems are still found today. An example of a system with an analogous method of forming its numbers comes from Papua New Guinea (Table 3.2). Waskia has a pair-counting system whose numbers are based on its terms for *one* and *two*:[6]

An example of counting by threes is also found in Papua New Guinea. The Bukiyip people speak an Arapesh language (Table 3.3) and name many of the groups of *three* after parts of the body:[7]

Bukiyip finger-counting reaches *six* on each hand (hence, the word for the number *six* is the word for *hand*) by using all five fingers and adding a thumb joint.[8] *Twelve* is reached by using both hands. A remote Arapesh-speaking group was reported to count to *twelve* differently: A first group of *three* was counted on the little, ring, and middle fingers of the first hand, a second group of *three* on the index finger and thumb of the first hand and the thumb of the second hand, and a third group of *three* on the index, middle, and ring fingers of the second hand; a final group of *three* was counted mentally, leaving the little finger of the second hand unused.[9] Counting beyond *twenty-four* involved using "a stick or a peg" to record each completed group before counting resumed.[10]

A so-called woman's dialect was possibly another minority tradition. Known as *eme-sal*, this dialect may have been a gendered variant of Sumerian; in literary texts, it was used in recording the utterances of women, goddesses, and eunuchs.[11] It contained just a few numbers (Table 3.4):

Today, gendered differences in spoken languages are found all over the world, with the possible exception of Africa.[12] An example is Japanese, where women might express the phrase *wait a moment* as *shoushou omachi kudasai* (a longer, more formal expression), whereas men might say *chotto matte* (a shorter, more abrupt form). Of course, both genders might use either form, but women using the shorter expression might be considered impolite, whereas men using the longer form would seem effeminate. Politeness is not the only path to developing a woman's dialect. Another is "dialect merger," which occurs when geographically distinct

[6] Ross and Paol (1978).

[7] Conrad and Wogiga (1991).

[8] Fortune (1942), p. 59.

[9] Fortune (1942), pp. 59–60.

[10] Fortune (1942), p. 60.

[11] Landsberger (1956); Whittaker (2002).

[12] Dunn (2014), p. 48.

Table 3.3 Bukiyip counting system (Conrad & Wogiga, 1991, pp. 74–76)

Nr.	Bukiyip number	Gloss	Nr.	Bukiyip number	Gloss
1	*ati-ny*	1	13	*wi-wis ati-ny*	*hand-hand*-1
2	*bie-ch*	2	14	*wi-wis biech*	*hand-hand*-2
3	*bie-ch ati-ny*	2 + 1	15	*aiag-ich*	*leg*
4	*nobat-ich*	4	16	*aiag ati-ny*	*leg*-1
5	*nobat ati-ny*	4 + 1	17	*aiag bie-ch*	*leg*-2
6	*onowip-ich*	*hand*	18	*élman-igú*	*man*
7	*onowip ati-ny*	6 + 1	19	*élman-igú ati-ny*	*man*-1
8	*onowip bie-ch*	6 + 2	20	*élman-igú bie-ch*	*man*-2
9	*anap-ich*	9	21	*én-ugú chikn-igú*	*some full*
10	*anap ati-ny*	9 + 1	22	*én-ugú chikn-igú ati-ny*	*some full*-1
11	*anap bie-ch*	9 + 2	23	*én-ugú chikn-igú bie-ch*	*some full*-2
12	*wi-wis-ich*	*hand-hand*	24	*élman chikni-nú*	*man full*

Table 3.4 Eme-sal and their Sumerian counterparts (Blažek, 1999, p. 329)

Nr.	Eme-sal	Sumerian
1	*de*	*diš*
2	*nim*	*min*
3	*am(m)uš*	*eš*
60	*mu-uš* or *min-eš*	*g̃eš*

dialects come together,[13] as can happen when women are exchanged between cultural groups as wives, concubines, servants, and slaves, a practice that was true of the ancient Near East. An example of dialect merger in numbers is found in the gendered pronunciation of *two* in Chukchi, a minority language spoken in Russia.[14]

Minority numerical traditions tend to disappear. Today, when minority traditions encounter systems that count to higher numbers, like Western numbers, the minority numbers tend to be replaced by the foreign numbers—except, perhaps, for the very smallest and oldest numbers: *one*, *two*, and *three*. There are neurological reasons for this: because quantities up to *about three or four* are perceptually appreciable without counting, "their mental representation … is much more precise, more deeply anchored in cognition, and therefore less susceptible to change," relative to quantities beyond that range.[15] This would have been true in Mesopotamia as well, as ancient peoples shared the same ability to perceive quantity.[16] Thus, it's interesting that the eme-sal numbers are *one*, *two*, and *three*, which are consistent with this phenomenon, and *sixty*, which is not.

[13] Dunn (2014), p. 60.

[14] Dunn (2018).

[15] Gardani and Zanini (2022), p. 264.

[16] Overmann (2015, 2019).

Foreign numbers are adopted for many reasons. An example is found in Polynesia, where the traditional numbers and counting practices persisted for thousands of years, including centuries of infrequent maritime contact with Westerners. Then missionaries arrived, bringing with them the specific intent to impose not just the Christian religion but Western education, social conduct, and dress on the Indigenous population.[17] Missionaries targeted the traditional number system, believing that decimal numbers as they were structured in the Western tradition were "more convenient."[18] Within just a few decades, the traditional Polynesian counting practices and multiple counting sequences had effectively been erased. Polynesians not only adopted standard decimal numbers and English words like *hundred* and *thousand*, dictionaries began listing only standard decimal versions of Polynesian numbers, as if these were all that had ever existed.

However, Polynesian peoples also actively embraced Western numbers, as attested by an 1842 letter to a Hawaiian language newspaper,[19] published about the time traditional Polynesian numbers were disappearing from dictionaries. Mastery of Western numbers was recognized as crucial if the *kama'aina* or *people of the land* were to be successful in negotiating economic, legal, and political matters with foreign states, and retain their sovereignty and cultural identity. Similar considerations would have once motivated the cultural groups in Mesopotamia—even those that were more dominant—to adopt the Sumerian numbers.

References

Anderson, R. (1864). *The Hawaiian Islands: Their progress and condition under missionary labors*. Gould and Lincoln.

Balke, T. E. (2010). The Sumerian ternary numeral system. *Journal of Cuneiform Studies, 62*, 45–52.

Blažek, V. (1999). *Numerals: Comparative etymological analyses and their implications: Saharan, Nubian, Egyptian, Berber, Kartvelian, Uralic, Altaic and Indo-European languages*. Masarykova Univerzita.

Civil, M. (2005). Nos. 61: Late Babylonian grammatical text. In I. Spar & W. G. Lambert (Eds.), *Cuneiform texts in the Metropolitan Museum of Art. Vol. II: Literary and scholastic texts of the first millennium B. C* (pp. 244–247). Brizzolis Arte en Gráficas.

Conrad, R. J., & Wogiga, K. (1991). *An outline of Bukiyip grammar*. Australian National University.

Diakonoff, I. M. (1983). Some reflections on numerals in Sumerian: Towards a history of mathematical speculation. *Journal of the American Oriental Society, 103*(1), 84–92.

Dunn, M. (2014). Gender determined dialect variation. In G. G. Corbett (Ed.), *The expression of gender* (pp. 39–68). Walter de Gruyter.

Dunn, M. (2018). Chukchi. In E. Chan (Ed.), *Numeral systems of the world's languages*. Max Planck Institute. https://lingweb.eva.mpg.de/channumerals/Chukchi.htm

[17] Anderson (1864).

[18] Hale (1846), p. 247.

[19] Lagaro (1842).

Eberhard, D. M., Simons, G. F., & Fenning, C. D. (Eds.). (2024). *Ethnologue: Languages of the world* (25th ed.). SIL International. https://www.ethnologue.com

Fortune, R. F. (1942). *Arapesh* (Vol. 6). J. J. Augustin.

Gardani, F., & Zanini, C. (2022). On the subitizing effect in language contact. In R. Levie, A. Bar-On, O. Ashkenazi, E. Dattner, & G. Brandes (Eds.), *Developing language and literacy: Studies in honor of Dorit Diskin Ravid* (pp. 263–293). Springer Nature.

Hale, H. (1846). *United States exploring expedition. During the years 1838, 1839, 1840, 1841, 1842. Under the command of Charles Wilkes, U.S.N. Vol. IV: Ethnography and philology*. C. Sherman.

Lagaro, N. (1842). Ka waiwai o ka helu [The value of the number]. *Ka Nonanona*, *2*(10), 46–47.

Lambert, W. G. (1969). Review, O. R. Gurney and P. Hulin, The Sultantepe tablets II. *Journal of Semitic Studies, 14*(2), 242–247.

Landsberger, B. (1956). Part I: Emesal-vocabulary (series dimir-dingir-ilum). In B. Landsberger, R. T. Hallock, T. P. R. Jacobsen, & A. Falkenstein (Eds.), *Materialien zum Sumerischen Lexikon [Materials for the Sumerian lexicon] (MSL IV)* (pp. 1–44). Pontificium Institutum Biblicum.

Overmann, K. A. (2015). Numerosity structures the expression of quantity in lexical numbers and grammatical number. *Current Anthropology, 56*(5), 638–653.

Overmann, K. A. (2019). *The material origin of numbers: Insights from the archaeology of the ancient Near East*. Gorgias Studies in the Ancient Near East 14. Gorgias Press.

Ross, M., & Paol, J. N. (1978). *A Waskia grammar sketch and vocabulary*. Australian National University.

Thureau-Dangin, F. (1928). Le système ternaire dans la numération sumérienne [The ternary system in Sumerian numeration]. *Revue d'Assyriologie et d'archéologie Orientale, 25*, 119–121.

Whittaker, G. (2002). Linguistic anthropology and the study of emesal as (a) women's language. In S. Parpola & R. M. Whiting (Eds.), *Proceedings of the 47e Rencontre Assyriologique Internationale, Helsinki, 2001: Sex and gender in the ancient Near East* (Part II, pp. 633–644). The Neo-Assyrian Text Corpus Project.

Chapter 4
Akkadian Numbers

Abstract Akkad was the land to the northwest of Sumer in what is today Syria. Although their indigenous numbers were decimal, the Akkadians borrowed the Sumerian sexagesimal numbers along with the technologies for representing them, tokens and written notations. Proto-languages, along with analyses of the Akkadian number-words, suggest that Akkadian numbers were quite old, and possibly even older than the Sumerian numbers. Scribes speaking Eblaite, a Semitic language related to Akkadian, were the first to record the Sumerian words for numbers, most likely as a way of learning the Sumerian number-words in addition to the Sumerian numerical notations (much like we might today be inspired to learn the Latin words for numbers to use with the Roman numerals). This development may have shown the Akkadians how to adapt the cuneiform writing system to the Akkadian language. Akkadian use would eventually influence the Sumerian numbers in dramatic ways, for example, changing them from sexagesimal to decimal and organizing them with place value. These changes were likely facilitated by the observable differences between Akkadian decimal and Sumerian sexagesimal numbers.

Keywords Proto-language · Analyzability · Etymological transparency · Eblaite · Assyro-Babylonian common

To the northwest of ancient Sumer lay Akkad, today known as Syria. The inhabitants spoke a now-extinct language related to Arabic and Hebrew. While the Akkadians had a decimal number system, they also adopted the Sumerian sexagesimal numbers and used them for several thousand years. Less well known are the ways in which they adapted these foreign numbers to suit their own, decimal, purposes. Because of these adaptations, Akkadian numbers had a significant impact on the numerical legacy of the ancient Near East, ultimately contributing decimal numbers with just a hint of sexagesimality to the contemporary Western numerical tradition.

Akkadian numbers were very old, and probably older than Sumerian numbers. In the early fourth millennium BCE, the reconstructed ancestor of the Akkadian

K. A. Overmann, *Cultural Number Systems*, Interdisciplinary Contributions to Archaeology, https://doi.org/10.1007/978-3-031-83383-0_4

language, Proto-Semitic, had words for *thousand* and *ten thousand*.[1] This tells us several things. First, Akkadian numbers were decimal, organized around the number *ten*. Second, the difference in structure—decimal for Akkadian, sexagesimal for Sumerian—means that the two numerical traditions originated independently and before the two societies came into much contact with each other in the Neolithic, and that the Akkadians were unlikely to have invented the sexagesimal system of tokens or numerical impressions, as scholars occasionally suggest.[2] Finally, counting into the thousands and tens of thousands means that Akkadian numbers were already old by the early fourth millennium BCE. This is because counting to such high numbers doesn't develop until a number system is well established, which can take thousands of years.

How old these number systems actually were, though, is a good question. Independence implies an origin before the Neolithic period (8300–4500 BCE) when they came into contact, and Proto-Semitic's ancestor, Proto-Afro-Asiatic, has a few small numbers, perhaps as old as 20,000 years.[3] And while both numerical traditions were old, the Akkadian tradition was likely the older of the two. In any numerical tradition, small numbers (for our purposes, *one* through *ten*) are the oldest because they emerge the earliest in any system of numbers. When small numbers are used over long periods of time, or when they are used frequently, the associated words become unanalyzable. An *unanalyzable* number is a word like *nine* that cannot be broken into smaller numbers like *nineteen* means *nine* [and] *ten*. This had already occurred for small Akkadian numbers,[4] but not for small Sumerian numbers, because the Sumerian words for *six* through *nine* are analyzable as *five* plus the appropriate smaller number.[5] This difference in analyzability, along with the frequent use of numbers in both cultural groups, suggests that Akkadian numbers are older than Sumerian numbers.

The Akkadians didn't simply adopt Sumerian numbers; they also adapted them. A particularly consequential adaptation was recording the Sumerian words for numbers. Numerals like "7" (Western) and "𒐌" (cuneiform) don't convey sound values in the way that written words like "*seven*" (English) and "*imin*" (Sumerian) do. The Sumerian numerals were used without their sound values being recorded for a period spanning several centuries (small numbers) to more than a thousand years (large numbers).[6] While some of this delay reflected the need to invent techniques for recording sound values, much of it occurred because numerical signs work better as numbers without sound values. This is because sound values make signs visually complex, and visual complexity degrades the conciseness that makes numerals

[1] Kitchen et al. (2009); Lipínski (2001).

[2] Englund (2006); Veldhuis (2014).

[3] Ehret (1995).

[4] Miller and Shipp (2014).

[5] Edzard (1980); Pettinato (1981).

[6] Damerow (1988).

able to display relations and patterns.[7] So, there had to be a reason to record the sound values of words for numbers.

A possible reason is suggested by a tablet known as TM.75.G.2198 (Fig. 4.1), one of the earliest records of the Sumerian words for numbers. This tablet gives the sound values for the Sumerian words *two* through *four* and *six* through *ten*; it lists the numerals 1 and 5 but doesn't phonetically specify the associated words.[8] The tablet was found in Ebla, a city west of Akkad whose inhabitants spoke Eblaite, a Semitic language related to Akkadian. By this point in time, Sumerian was a classical language, spoken only by scribes and scholars. The scribe who wrote the tablet likely wanted to learn the Sumerian words for numbers, in addition to using the sexagesimal notations. A modern parallel would be our learning the Latin words for Roman numerals. Unless we happen to be studying Latin, most of us don't bother, since the Roman numerals are perfectly usable without the Latin vocabulary. We simply use Roman numerals while thinking and speaking our own number system and language. This is what ancient peoples borrowing the Sumerian numerals would have done as well, and it explains why they were able to use the foreign numerals—even for several thousand years—without their own numbers being replaced.

The tablet was written in the late third millennium BCE, just before the cuneiform script was adapted to write the Akkadian language, which of course had different sound values than Sumerian did. The timing suggests that writing the sound values of the Sumerian number-words was an initial means and motivation for writing the sounds and meanings of Akkadian.[9]

Another consequential adaptation was *place value*, which emerged around the time that cuneiform was being adapted to Akkadian.[10] Place value drew upon characteristics that were used in earlier forms, like linearity (aligning numbers horizontally) and magnitude ordering (sequencing numbers from largest to smallest), as

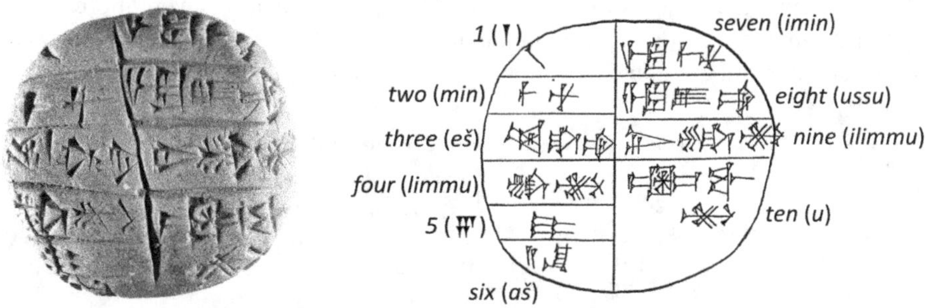

Fig. 4.1 Lexical tablet (TM.75.G.2198) from the city of Ebla dated to 2350–2250 BCE. (Left) Photograph by M. Necci and (right) line drawing from Edzard (1980, Figs 26a and 26b). Copyright Missione Archeologica Italiana in Siria and used by permission. Sumerian words for numbers from Blažek (1999). A version was previously published in Overmann (2023, Fig. 13.8, p. 329)

[7] Overmann (2022).

[8] Edzard (1980); Friberg (1986); Pettinato (1981).

[9] Overmann (2021).

[10] Høyrup (2002); Robson (2007).

well as practices that aligned written notations (vertical columns arranged by shared units). Reformulating these structural elements as place value would also have leveraged insights that Akkadian scribes acquired from the contrast between decimal and sexagesimal as organizing bases. A modern parallel would be our noticing that Roman numerals have a subbase of *five*—as in the single signs for 5 (V), 50 (L), and 500 (D)—and using that insight to reconfigure Western numerals. The contrast in structure—Akkadian decimal, Sumerian sexagesimal—made it more likely that the base organization would have been noticed and developed.

Decimal numbers became more prevalent than sexagesimal numbers. Today, the Sumerian sexagesimal numbers are famous as the number system used in Mesopotamia. However, by the beginning of the second millennium BCE, a decimal system of notations, Assyro-Babylonian Common (ABC; see Fig. 4.2), had come into widespread use, reflecting the increasing cultural influence of Semitic peoples like the Akkadians and Eblaites.[11] The signs used, and a special sign for *sixty*, show that ABC decimal was adapted from the sexagesimal system, specifically the cuneiform notations. But once it was available, ABC decimal wasn't simply used alongside the sexagesimal system: By the second millennium BCE, it was the more predominant of the two, despite not being as well known today.[12] It was this system that would eventually contribute decimal numbers—with a unique and lingering hint of sexagesimal in the 360 degrees that make a circle—to the Western numerical tradition.[13]

The Akkadians also changed the sexagesimal numerals. In the first millennium BCE, the way that large numbers were written in the sexagesimal system was changed in a way that reflects the influence of decimal thinking (Fig. 4.3). In the fourth millennium BCE with tokens, and in the third millennium BCE with proto-cuneiform, the sign for 36,000 was a small sphere (the sign for 10) superimposed over a large sphere (the sign for 3600). In the first millennium BCE, this complex sign was separated as signs for 10 and 3600.[14] The new notations were more reflective of decimal structure, suggesting the influence of decimal thinking on borrowed sexagesimal numbers. The new notations also represented explicit relations of

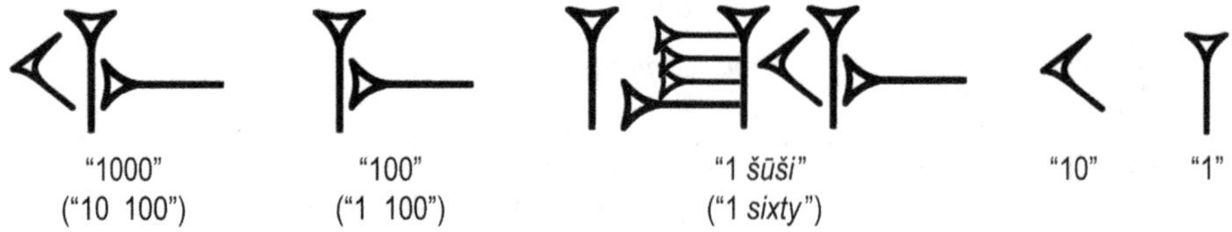

Fig. 4.2 Assyro-Babylonian Common, a decimal system in widespread use by the beginning of the second millennium BCE. Image by the author based on data from Chrisomalis (2010, Table 7.13, p. 248)

[11] De Odorico (1995); Thureau-Dangin (1939).

[12] Chrisomalis (2010).

[13] Joseph (2011).

[14] Powell (1971).

4th mil. BCE "36,000 × 1" | 3rd mil. BCE "36,000 × 1" | 1st mil. BCE "3600 × 10"

Fig. 4.3 The representation of 36,000 changed in the first millennium BCE. Image by the author. The signs are Unicode characters

multiplication that had developed between the fourth and first millennia BCE,[15] as well as the waning of Sumerian cultural influence within the Mesopotamian region.

References

Blažek, V. (1999). *Numerals: Comparative etymological analyses and their implications: Saharan, Nubian, Egyptian, Berber, Kartvelian, Uralic, Altaic and Indo-European languages*. Masarykova Univerzita.

Chrisomalis, S. (2010). *Numerical notation: A comparative history*. Cambridge University Press.

Damerow, P. (1988). Individual development and cultural evolution of arithmetical thinking. In S. Strauss (Ed.), *Ontogeny, phylogeny, and historical development: The Tel Aviv annual workshop in human development* (Vol. 2, pp. 125–152). Ablex Publishing Corporation.

De Odorico, M. (1995). *The use of numbers and quantifications in the Assyrian royal inscriptions*. University of Helsinki.

Edzard, D. O. (1980). Sumerisch 1 bis 10 in Ebla [Sumerian 1 to 10 in Ebla]. *Studi Eblaiti, III*(5–8), 121–127.

Ehret, C. (1995). *Reconstructing Proto-Afroasiatic (Proto-Afrasian): Vowels, tone, consonants, and vocabulary*. University of California Press.

Englund, R. K. (2006). An examination of the "textual" witnesses to the Late Uruk world systems. In Y. Gong & Y. Chen (Eds.), *Special issue of oriental studies: A collection of papers on ancient civilizations of Western Asia, Asia Minor and North Africa* (pp. 1–38). Oriental Studies.

Friberg, J. (1986). The early roots of Babylonian mathematics III. Three remarkable texts from ancient Ebla. *Vicino Oriente, 6*, 3–25.

Høyrup, J. (2002). A note on Old Babylonian computational techniques. *Historia Mathematica, 29*(2), 193–198.

Joseph, G. G. (2011). *The crest of the peacock: Non-European roots of mathematics* (3rd ed.). Princeton University Press.

Kitchen, A., Ehret, C., Assefa, S., & Mulligan, C. J. (2009). Bayesian phylogenetic analysis of Semitic languages identifies an Early Bronze Age origin of Semitic in the Near East. *Proceedings of the Royal Society of London. Series B, Biological Sciences, 276*(1668), 2703–2710.

Lipínski, E. (2001). Numerals. In *Semitic languages: Outline of a comparative grammar* (2nd ed., pp. 288–313). Peeters.

Miller, D. B., & Shipp, M. R. (2014). *An Akkadian handbook: Helps, paradigms, glossary, logograms, and sign list* (2nd ed). Eisenbrauns.

[15] Robson (2007, 2008).

Overmann, K. A. (2021). Writing system transmission and change: A neurofunctional perspective. In G. Gabriel, K. A. Overmann, & A. Payne (Eds.), *Signs – sounds – semantics. Nature and transformation of writing systems in the ancient Near East* (pp. 93–116). Ugarit-Verlag.

Overmann, K. A. (2022). Early writing: A cognitive archaeological perspective on literacy and numeracy. *Visible Language, 56*(1), 8–44.

Overmann, K. A. (2023). *The materiality of numbers: Emergence and elaboration from prehistory to present*. Cambridge University Press.

Pettinato, G. (1981). La pronuncia sumerica dei numeri da 1 a 10 in un testo lessicale di Ebla [The Sumerian pronunciation of the numbers 1 to 10 in a lexical text from Ebla]. *Annali Istituto Orientale Di Napoli Roma, 41*(1), 141–143.

Powell, M. A. J. (1971). *Sumerian numeration and metrology.* Unpublished doctoral dissertation, University of Minnesota.

Robson, E. (2007). Mesopotamian mathematics. In V. Katz (Ed.), *The mathematics of Egypt, Mesopotamia, China, India, and Islam: A sourcebook* (pp. 57–186). Princeton University Press.

Robson, E. (2008). *Mathematics in ancient Iraq: A social history*. Princeton University Press.

Thureau-Dangin, F. (1939). Sketch of a history of the sexagesimal system. *Osiris, 7*, 95–141.

Veldhuis, N. (2014). *History of the cuneiform lexical tradition*. Ugarit-Verlag.

Chapter 5
Elamite Numbers

Abstract Elam was the land to northeast of Sumer and east of Akkad in what is today Iran. Very little is known about the Elamite number system beyond the fact that, like the Akkadians, they borrowed the Sumerian sexagesimal numbers along with the technologies for representing them, tokens and written notations. Like the Akkadians, the Elamites also adapted the Sumerian numbers they borrowed, for example, developing a unique decimal system for counting people and sheep. The Elamite decimal numbers were possibly used in paying wages to local workers, who would not have known the Sumerian sexagesimal numbers used by scribes. A few mysterious notations suggest that the Elamite decimal numbers may have been imported into Sumer, but they don't seem to have caught on there, possibly because the number system was short-lived in Elam itself. As decimal systems generally emerge from counting on the fingers to *ten*, the Elamite decimal numbers probably emerged from finger-counting. However, since the Elamite words for numbers were never recorded and their finger-counting was never described, there is no evidence for how the Elamites might have counted on their fingers.

Keywords Linguistic isolate · Animacy · Social value · Finger-counting · Mental effort

Northeast of Sumer and east of Akkad lay the land of Elam, today Iran. The region was inhabited by a people whose language not only went extinct, it was also an isolate, unrelated to any other known language, living or dead. Not very much is known about the Elamite number system, since there aren't any related languages to provide insight like there is for Akkadian. In addition, the Elamite words for numbers were never recorded,[1] like the Sumerian number-words weren't until other societies had both the means and motivation for doing so. Like the Akkadians, the Elamites too adopted and used the Sumerian tokens and proto-cuneiform notations. Uniquely, the Elamites also created decimal versions of these early counting

[1] Tavernier (2020).

K. A. Overmann, *Cultural Number Systems*, Interdisciplinary Contributions to Archaeology, https://doi.org/10.1007/978-3-031-83383-0_5

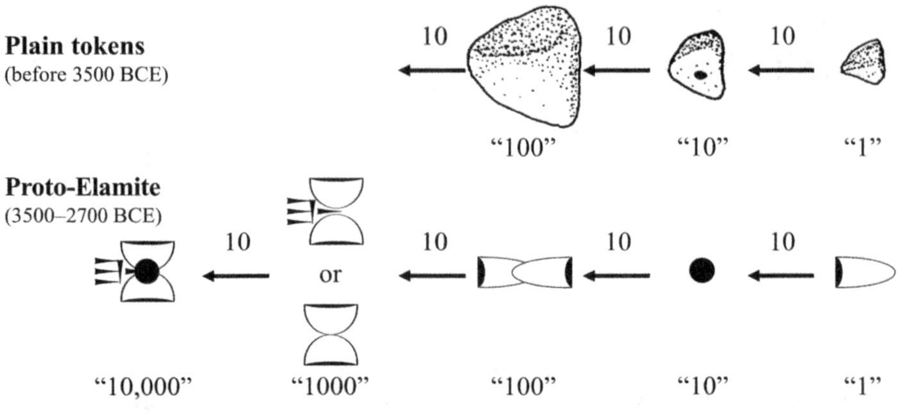

Fig. 5.1 Elamite decimal tokens and proto-Elamite notations. Tokens (top) adapted from Friberg (2019, p. 186); proto-Elamite (bottom) redrawn from Englund (2004, Fig. 5.4, p. 107)

technologies (Fig. 5.1),[2] and this implies that their own numbers were decimal. It's possible that the Elamites created decimal versions for the purpose of trading with the Akkadians, whose own numbers were strongly and persistently decimal. However, there doesn't seem to be any trace of these systems in Akkad, which suggests that these systems were most likely created for purposes internal to Elam.

The Elamites transformed the adopted technologies to reflect their own numbers. When societies adopt foreign technologies like tokens and numerical notations, they are also likely to adapt them by making them more like what they already know because this makes them more familiar and easier to use. Accordingly, when the Elamites adapted the Sumerian counting technologies, they did so in ways that reflected their own number system.

The Elamites used decimal versions of tokens and proto-notations to count discrete animate objects, particularly human laborers and herd animals like sheep.[3] Animacy is a grammatical and semantic feature found in some languages, including Elamite. It's an implicit hierarchy of value that runs from deities and people through domesticated animals to inanimate objects.[4] Using the decimal versions to count animate objects implies that the objects being counted were socially valuable in some way, suggesting a possible motive for creating the decimal notations: Something socially important was being counted with the society's own numbers. The decimal system might also have been created with a particular audience in mind: laborers and shepherds. These workers were less likely to have acquired foreign numbers through education. Accounting for their labor in the more familiar

[2] Englund (1996); Friberg (2007, 2019).

[3] Englund (1996), p. 162.

[4] Yamamoto (1999), p. 1; also see Sorlin and Gardelle (2018).

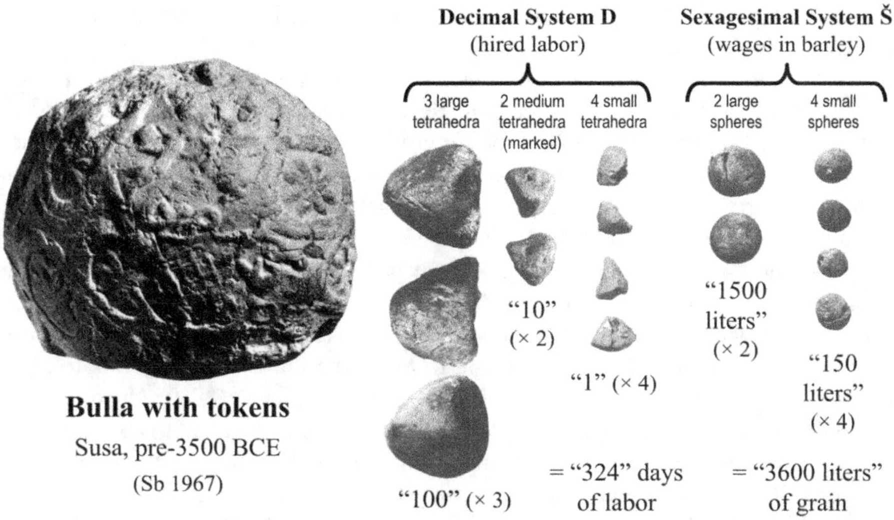

Fig. 5.2 Bulla with tokens (Sb 1967) from the city of Susa, dated to before 3500 BCE. The tokens show the Elamite decimal system being used alongside the sexagesimal system for barley. Adapted and labeled from Amiet (1972, entry 488)

Indigenous numbers might have made the payment process easier and more transparent for all concerned.

While it was used, the Elamite decimal system appeared alongside other, sexagesimal, counting sequences. An example is shown by an early bulla from the city of Susa (Fig. 5.2), whose associated tokens have been interpreted as the decimal system for hired labor and the sexagesimal system for barley.

Elamite decimal numbers may have reached Sumer, but simply failed to catch on. While the proto-Elamite decimal system had "no parallel" in proto-cuneiform, five tablets found in Sumerian cities contain numerical signs that are "possible but unlikely exceptions."[5] These tablets contain proto-cuneiform notations that curiously resemble the proto-Elamite sign for 100 (see Fig. 5.1, bottom row). The tablets were found in the cities of Uruk and Umma, and they have been dated to between 3500 and 3000 BCE. It's possible that these notations represent an as-yet unidentified counting sequence from the Sumerian numerical tradition, which, as was noted in the chapter on Sumerian numbers (Chap. 2), had more than a dozen counting sequences for different types of objects.[6] Alternatively, they might have been an intrusion of the Elamite decimal tradition into neighboring Sumer, perhaps as transported there by Elamite merchants.

The Elamite decimal system didn't last very long. It appeared around the mid-fourth millennium BCE, about the same time that proto-cuneiform writing first

[5] Englund (2004), p. 108.

[6] Chambon (2003); Damerow and Englund (1989); Englund (2004).

emerged in Sumer. It disappeared, along with the short-lived proto-Elamite script, after just a few centuries.[7]

Elamite decimal structure most likely emerged from finger-counting. There's really no way to tell, after-the-fact, how decimal structure emerged, except to say that it's based on using the fingers for counting. However, the Elamites likely used one of the several methods found around the world today.[8] One such method is known in North America: With a pair of hands acting as a tally that can count to *ten*, the numerical extent of the tally is increased by recruiting and using the hands of more people.[9] The first person counts from *one* to *ten*, the second person counts from *eleven* to *twenty*, the third from *twenty-one* to *thirty*, and so on. In a vigesimal (or base-20) system, the toes would also be used, so the first person counts from *one* to *twenty*, the next one from *twenty-one* to *forty*, the third from *forty-one* to *sixty*, and so on.[10] Counting remains simple—an advantage because it reduces mental effort—since each person just counts on his or her digits from *one* to *ten* (or *twenty* if the toes are used). The organization of Elamite numbers around the number *ten* suggests that they used their hands for counting and didn't use their feet.

Another possible method the Elamites might have used, one found in Oceania and Africa, also involves cooperation between different people. Instead of adding numerical extent to the tally, each person keeps track of a different register: units, tens, hundreds, thousands, and so on.[11] In this method, each person again counts only from *one* to *ten*. When the person keeping track of the units reaches *ten*, the person counting the tens counts *one*. Similarly, when the person keeping track of the tens reaches *ten*, the person counting the hundreds counts *one*, and so on.

A third counting strategy is found in Papua New Guinea. A man counts to *twenty* on his fingers and toes, and then imagines that each of his digits accumulates the *twenty* digits of a person, which he can keep track of with his own digits; in this way, a single individual can count to 400 (see Fig. 27.3).[12] A decimal variant of this same method is suggested by certain formations in North American numbers that imply the person counting keeps track of the cycles of *ten* in some fashion.[13]

What all these methods have in common is that they minimize the amount of mental effort by counting only to *ten* (or *twenty*) and using the fingers (and perhaps the toes) to keep track of the count. Whatever method of counting the Elamites actually used, it too would have reflected these universal tendencies to keep things simple and make counting efficient.

[7] Friberg (2019).

[8] Overmann (2021).

[9] Kroeber and Grace (1960); Sparkman (1905).

[10] Nansen (1893).

[11] Cajori (1899); Collocott (1925, 1927); Schrumpf (1862).

[12] Mimica (1988).

[13] Overmann (2021).

References

Amiet, P. (1972). *Mémoires de la délégation archéologique en Iran, Tome XLIII, Mission de Susiane [Memoirs of the archaeological delegation to Iran, Vol. XLIII, Mission of Susiane]* (Vol. I). Librairie Orientaliste Paul Geuthner.

Cajori, F. (1899). The number concept. *Colorado College Studies: Papers Read before the Colorado College Scientific Society, 8*, 32–40.

Chambon, G. (2003). Archaic metrological systems from Ur. *Cuneiform Digital Library Journal, 5*, 1–8.

Collocott, E. E. V. (1925). Supplementary Tongan vocabulary; also notes on measuring and counting, proverbial expressions and phases of the moon. *Journal of the Polynesian Society, 34*(3), 193–213.

Collocott, E. E. V. (1927). Kava ceremonial in Tonga. *Journal of the Polynesian Society, 36*(1), 21–47.

Damerow, P., & Englund, R. K. (1989). *The proto-Elamite texts from Tepe Yahya*. Harvard University Press.

Englund, R. K. (1996). The proto-Elamite script. In P. T. Daniels & W. Bright (Eds.), *The world's writing systems* (pp. 160–164). Oxford University Press.

Englund, R. K. (2004). The state of decipherment of proto-Elamite. In S. D. Houston (Ed.), *The first writing: Script invention as history and process* (pp. 100–149). Cambridge University Press.

Friberg, J. (2007). *A remarkable collection of Babylonian mathematical texts*. Springer Science+Business Media, LLC.

Friberg, J. (2019). Three thousand years of sexagesimal numbers in Mesopotamian mathematical texts. *Archive for History of Exact Sciences, 73*, 183–216.

Kroeber, A. L., & Grace, G. W. (1960). Numerals. In *The Sparkman grammar of Luiseño* (pp. 118–120). University of California Press.

Mimica, J. (1988). *Intimations of infinity: The mythopoeia of the Iqwaye counting system and number*. Berg.

Nansen, F. (1893). *Eskimo life* (William Archer, Trans.). Longmans, Green, and Co.

Overmann, K. A. (2021). Finger-counting and numerical structure. *Frontiers in Psychology, 12*, 723492.

Schrumpf, C. (1862). Sessuto. Ein Beitrag zur Süd-Afrikanischen Sprachenkunde, geschöpft aus meiner sechszehnjährigen Erfahrung [Sessuto. A contribution to South African linguistics, drawn from my sixteen years of experience]. *Zeitschrift der Deutschen Morgenländischen Gesellschaft, 16*(3), 448–481.

Sorlin, S., & Gardelle, L. (2018). Anthropocentrism, egocentrism and the notion of animacy hierarchy. *International Journal of Language and Culture, 5*(2), 113–162.

Sparkman, P. S. (1905). Sketch of the grammar of the Luiseño language of California. *American Anthropologist, New Series, 7*(4), 656–662.

Tavernier, J. (2020). Elamite. In R. Hasselbach-Andee (Ed.), *A companion to ancient Near Eastern languages* (pp. 163–184). Wiley.

Yamamoto, M. (1999). *Animacy and reference: A cognitive approach to corpus linguistics*. John Benjamins.

Chapter 6
Numbers in the Achaemenid Empire

Abstract On his way to fight the Battle of Thermopylae, the fourth Persian King, Xerxes I, stopped in Thrace to count the men of his army. Xerxes' army was quite large, as his infantry alone was said to have contained 1.7 million men. The king chose a particularly awkward method for counting his men: he created an enclosure that could contain ten thousand of them at a time and then marched them all through it until they had all been counted. Over a century later, another Persian king, Darius III, used the same method to count the men in his army. Ever since the Greek historian Herodotus recorded the first incident with Xerxes I, scholars have tried to understand the reason for this unusual method of counting. Scholars have speculated that the method chosen represented a difficulty with large numbers. However, both Persian kings had at their disposal several number systems able to record large numbers. These included the number systems of Mesopotamia and Egypt, both of which were well-established and capable of counting into the millions. Most likely the method amounted to a kingly boast, something along the lines of, "my army is so big, I must count the men like grain!"

Keywords Xerxes I · Darius III · Numerical competence · Place value · Compactness

By the middle of the first millennium BCE, Elam and greater Mesopotamia and pretty much all of the ancient Near East and then some had been conquered by Cyrus the Great (ca. 600–530 BCE), who established the Achaemenid or First Persian Empire in 550 BCE. For its time, it was the greatest empire the world had ever seen, encompassing more than 2 million square miles of territory.[1] Less than a century later, the fourth Persian king, Xerxes I (ca. 518–465 BCE), was mounting his second invasion of Greece, where he would fight the famous Battle of

[1] Taagepera (1979); Turchin et al. (2006).

A shorter version of this chapter originally appeared in Overmann (2023), and the material is reused here with permission.

K. A. Overmann, *Cultural Number Systems*, Interdisciplinary Contributions to Archaeology, https://doi.org/10.1007/978-3-031-83383-0_6

Thermopylae (480 BCE). But first, in an exceedingly odd footnote to history, Xerxes apparently needed to count his men. Accordingly, Xerxes halted his army in a vast coastal plain in Thrace, a region that today overlaps the modern countries of Bulgaria, Turkey, and Greece. There, he had his men march, ten thousand at a time, into a circled enclosure that could contain that many men, assuming they were tightly packed.[2] Now, Xerxes supposedly mustered 1.7 million men in just his infantry alone. Assuming that it took a single hour to count each group—and ignoring the fact that an hour would not be nearly enough time to organize and move ten thousand people, no matter how loosely packed—Xerxes would've needed a full week, working 24 h each day, to count his infantry with this unusual method. Further, there was more to his army than the infantry, so a week's worth of time would've been just the start.

Ever since the Greek historian Herodotus (c. 484–425 BCE) memorialized this incident, scholars have expressed disbelief. They have doubted the infantry's size, and they've scoffed at the notion that Xerxes wouldn't have known how big his army really was unless he first halted and counted the men.[3] Some have highlighted the inordinate amount of time that would have been required, along with its negative effects on the military enterprise: An army standing around waiting to be counted isn't fighting, and in the meantime, it still needs to eat.[4] A few have focused on possible reasons for the exaggeration: Perhaps the number was meant to flatter the king,[5] or frighten his enemies by reporting an army so massive. One scholar, observing that the enumeration process penned up the men like so many sheep for slaughter, read the passage as a sad commentary on the pathos of war.[6] Another suggested that Herodotus had made a mistake in translation, confusing *chiliad*, the Persian unit for a thousand men, with *myriad*, the Greek unit for ten times that number.[7]

The idea that Herodotus simply mistranslated is almost certainly wrong, since more than a century later, another Persian king, Darius III (ca. 380–330 BCE), was said by the Roman historian Quintus Curtius Rufus to have counted his men in exactly the same way. Darius reportedly wanted to show off his strength to bolster his men's confidence before an immanent battle.[8] This time, the counting procedure is said to have taken place from dawn to dusk, significantly less than the 24 h per day that we previously granted Xerxes. Darius' total force was also much smaller, only about a fifth of the size of Xerxes' infantry. Even so, Darius would have needed to repeat the procedure more than 30 times, and it still remains a particularly awkward and unusual way to count people.

---

[2] Herodotus 7.60, translated in Strassler (2007), p. 523.

[3] Macan (1908b), p. 155.

[4] Macan (1908a), p. 82; Flower (2007), p. 819.

[5] Grote (1851), p. 36.

[6] Crosby (1860), p. 271, note 2.

[7] Flower (2007), p. 821.

[8] Quintus Curtius 3.2–3, translated in Rolfe (1946), p. 73.

The awkward method of counting likely served a very specific purpose. Scholars have long interpreted the incident as meaning that ancient peoples were simply unable to comprehend very high numbers.[9] Yet the number systems these kings would have known could count into the millions and higher, and they had become complex mathematical traditions more than a thousand years earlier. Thus, numerical incompetence is hardly likely. This would be especially true for men like kings who were highly educated and routinely dealt with matters of trade and military logistics, endeavors for which the ability to comprehend numbers had been crucial for an even longer period of time. Let's grant that the ancient armies may have lacked the ability to report their strength with the desirable accuracy and alacrity;[10] it still doesn't follow that they would resort to such crude measures in order to count their men.

Let's grant instead that ancient peoples were skilled and efficient at counting physical items of any sort, including soldiers. On this supposition, a new explanation appears: Xerxes and Darius had so many men, they couldn't count them in the usual way, but instead had to measure them as if they were grain—by the number times the men could fill a container capable of holding ten thousand at a time.[11] This would be consistent with the other boasts Xerxes is reported to have made: His men were so numerous, they could drink a river dry; his men were so numerous, they would bankrupt in a day any city that tried to feed them.[12] Our failure to see this footnote to history as more of the same royal puffery shows just how far divorced our Western numbers have become from the practical realities of counting physical objects.

Xerxes had several number systems to choose from. Aramaic, a Western Semitic language related to the Eastern Semitic languages Akkadian and Eblaite, was the *lingua franca* of the Persian Empire in the first millennium BCE,[13] and the associated decimal number system (Fig. 6.1) was a descendent of both Assyro-Babylonian

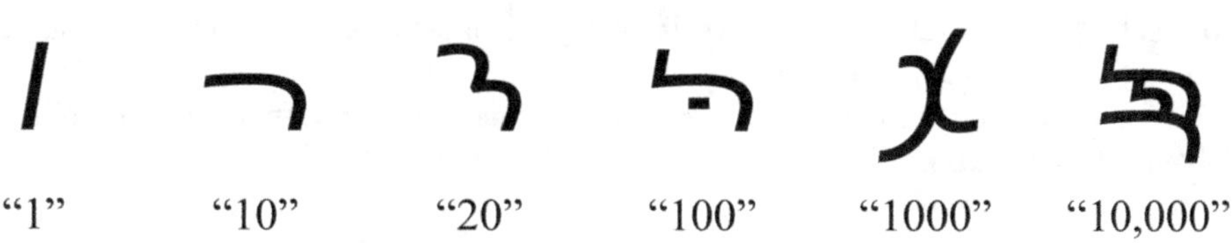

Fig. 6.1 Aramaic numerals in the first millennium BCE. As written, these numbers were ordered from right to left, the opposite direction used in Western numerals. For example, the number 55,874 would be written [illegible]. In order from largest to smallest (read from right to left), this is *five* ones and the sign for 10,000; *five* ones and the sign for 1000; and three signs for *twenty*, one sign for *ten*, and *four* ones. Adapted from Chrisomalis (2010, Tables 3.1 and 3.2, pp. 69–70)

[9] Flower (2007), p. 823.

[10] Maurice (1930), pp. 233–234, note 54.

[11] Macan (1908a), p. 82, note 4(i).

[12] Flower (2007), p. 819.

[13] Jacobs (2015).

no (10) and 7 (1)	5 (10) and 2 (1)	1 (10) and 3 (1)	2 (10) and no (1)
(7 × $\mathbf{60}^{\mathbf{3}}$ [216,000])	(52 × $\mathbf{60}^{\mathbf{2}}$ [3600])	(13 × $\mathbf{60}^{\mathbf{1}}$ [60])	(20 × $\mathbf{60}^{\mathbf{0}}$ [1])
1,512,000	187,200	780	20

Fig. 6.2 The number 1,700,000 in the sexagesimal cuneiform numerals with place value. The second row transliterates the numerals and units into Western notation. The third row shows place value (with the sexagesimal powers in bold); the numbers in brackets are the sexagesimal powers carried out. The comparable decimal exponents (from right to left) would be 10^0 (or 1, as any number raised to the zero power is 1), 10^1 (or 10, as any number raised to the first power is itself), 10^2 (or 100, as any number squared is multiplied by itself), and 10^3 (or 1000, as any number cubed is multiplied by itself twice). The bottom row shows the resultant multiplication. The final sum, 1.7 million, is achieved by adding the bottom row of numbers. Image by the author

Common (ABC; Fig. 4.2) and Egyptian hieroglyphic numbers.[14] Yet Xerxes might not have preferred the Aramaic numerals, as the highest symbol known—a sign for 10,000—would have required 170 repetitions, one for each group of ten thousand men measured with his boastful container. The ABC numerals were even more limited in this regard: Since the highest symbol was a thousand, 1700 repetitions would have been required. The difficulty of counting repetitions increases with the number of them, so while the ABC numerals were the most challenging in this regard, the Aramaic numerals would still have presented a problem.

ABC and Egyptian numbers, at least, were both decimal, and thus they had the advantage of being consistent with his supposed method of counting. In comparison, the sexagesimal system would have meant putting the decimally counted sums through some rather cumbersome conversions. However, the sexagesimal numbers had some other benefits. One was *place value*, which made these numbers—theoretically, at least—extensible to infinity, or as near to it a number as a king might try to reach. Another was *compactness*, in being six times as compressed as the same amount in decimal. As a result, the 1.7 million men of Xerxes' infantry could have been represented sexagesimally with just a handful of cuneiform numerals: 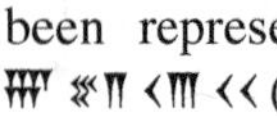 (Fig. 6.2).

References

Chrisomalis, S. (2010). *Numerical notation: A comparative history*. Cambridge University Press.

Crosby, W. H. (1860). *Quintus Curtius Rufus: Life and exploits of Alexander the Great* (2nd ed.). D. Appleton and Company.

Flower, M. A. (2007). The size of Xerxes' expeditionary force. In R. B. Strassler (Ed.), *The landmark Herodotus: The histories* (pp. 819–823). Pantheon Books.

Grote, G. (1851). *History of Greece* (Vol. V). John P. Jewett and Company.

[14] Chrisomalis (2010).

Jacobs, B. (2015). Achaemenid royal communication. In *Encyclopædia Iranica* (online ed.). https://iranicaonline.org/articles/achaemenid-royal-communication

Macan, R. W. (1908a). *Herodotus: The seventh, eighth & ninth books. Vol. I, Part 1: Introduction, Book VII (text and commentaries).* Macmillan.

Macan, R. W. (1908b). *Herodotus: The seventh, eighth & ninth books. Vol. II: Appendices, indices, maps.* Macmillan.

Maurice, F. (1930). The size of the army of Xerxes in the invasion of Greece 480 B.C. *Journal of Hellenic Studies, 50*(2), 210–235.

Overmann, K. A. (2023). *The materiality of numbers: Emergence and elaboration from prehistory to present.* Cambridge University Press.

Rolfe, J. C. (1946). *Quintus Curtius* (Vol. I). Harvard University Press.

Strassler, R. B. (Ed.). (2007). *The landmark Herodotus: The histories* (A. L. Purvis, Trans.). Pantheon Books.

Taagepera, R. (1979). Size and duration of empires: Growth-decline curves, 600 B.C. to 600 A.D. *Social Science History, 3*(3–4), 115–138.

Turchin, P., Adams, J. M., & Hall, T. D. (2006). East-west orientation of historical empires and modern states. *Journal of World-Systems Research, 12*(2), 219–229.

Chapter 7
Egyptian Numbers

Abstract Ancient Egypt was the civilization that formed around the Nile River. The Egyptians developed a distinctive system of writing, along with a system of decimal numbers. Egyptian counting on the fingers is mentioned in the *Book of the Dead*, which was written around 1600 BCE but was based on texts perhaps a thousand years older. The Greek historian Herodotus described the Egyptians as using pebbles to calculate, a form of abacus whose results were written in one of three scripts: small pictures and conventions known as *hieroglyphs*; a cursive form of hieroglyphs called *hieratic*; and a more abstract form named *demotic*. In numbers, the transition from hieroglyphs to hieratic to demotic forms shows how conventional forms for numbers (that is, signs for numbers whose meaning cannot be ascertained by counting the number of elements in the way a collection of vertical strokes can be counted) can emerge from handwriting and perceptual effects. One such perceptual effect influences the form of small numbers (*one* through *three*, consisting of one, two, or three vertical strokes) to be conserved, while higher numbers (*four* and above) change away from countable elements.

Keywords Number sense · Hieroglyphs · Cursive · Pebbles · Finger-counting

Before we leave the Near East, let's take a look at Egypt, the African gateway to the Near East and the ancient civilization that formed around the Nile River. Like its neighbors to the east, Egypt developed complex systems of numbers and writing. Scholars generally credit Mesopotamia with being slightly earlier in this regard, but it's difficult for us to know who was really the first because of the close proximity in time and space, the vagaries of what survives archaeologically and textually over thousands of years, and the imprecision of archaeological dating. It's also difficult to say that Egyptian and Mesopotamian numbers and writing were invented wholly independently of one another, given their history of contact and the likelihood of cultural diffusion. Nonetheless, even if the ideas of numbers and writing were shared between the two regions, these cultural systems developed uniquely, suggesting their independence. There's no hint of sexagesimal, for example, in Egypt's decimal numbers.

K. A. Overmann, *Cultural Number Systems*, Interdisciplinary Contributions to Archaeology, https://doi.org/10.1007/978-3-031-83383-0_7

Egyptians counted on their fingers. The Egyptian *Book of the Dead* was written around 1600 BCE, based on texts thought to be about a thousand years older. It describes a spell in which the fingers are used to count to *ten*: When the boat used to ferry the recently deceased into the underworld is summoned, the ferryman says he would be asked by the god of the underworld, "Did you bring me a man who cannot number his fingers?"[1] To this query, the supplicant replies, "I know how to count my fingers; take one, take the second, quench it, remove it, give it to me, be friendly towards me; do not let go of it; have no pity on it; make the Eye bright; give the Eye to me."[2]

Egyptians used pebbles to calculate. In his *Histories*, Herodotus often described foreign customs and noted many points of difference from Greek habits. After describing particularities of Egyptian climate, gendered employment, carrying burdens, urination and defecation, eating, religious employment, haircuts and grooming, mourning, keeping pet animals, brick-making, circumcision, clothing, and boat sails, Herodotus says that "in writing characters and numbers, the Greeks write from left-to-right, the Egyptians write from right-to-left."[3] At issue is the direction in which numerical notations were written, although in actuality, Egyptian numbers could be and were written in either direction. The same passage has also been translated as "in writing characters and *reckoning with pebbles*, the Greeks write from left-to-right, the Egyptians write from right-to-left" (emphasis added).[4] Since the ancient passage admits some modern interpretation, the second translation takes the view that *ψῆφος* meant the pebbles used for numbering, calculating and voting.[5] Since written notations are fixed, it make senses that the ancient Egyptians would have used a manipulable form like pebbles for calculating, creating a kind of abacus.

Egyptian numbers were written in multiple scripts. Today, hieroglyphs (3250 BCE–400 CE) are probably the most well-known of the Egyptian writing systems. These signs—small images of people, animals, and objects; conventions with learned social meanings[6]—maintained their ideal forms throughout the lifespan of the writing system, most likely because ideal forms honored the gods and kings. The hieroglyph numerals are shown in Fig. 7.1. Hieratic, a cursive or somewhat sloppier but more quickly written form of hieroglyphs, emerged about six centuries later (2600 BCE); once available, hieratic was used alongside hieroglyphs for their duration.[7] Demotic, a cursive whose character forms were fairly removed from the original hieroglyphs, emerged early in the first millennium BCE. Hieratic and demotic forms provide an opportunity to discuss two things: how written forms change over time, and how numerals differ from other writing.

[1] Neugebauer (1969), p. 9.

[2] Von Dassow (1998), p. 126.

[3] Herodotus II: 36.4; see translations in Godley (1922); Rawlinson (1875).

[4] Smith (1925).

[5] Personal communication from Antonis Iliopoulos dated September 4, 2019.

[6] Gardiner (1950); Imhausen (2016); Menninger (1992).

[7] Hoffmann (2012); Lopriano (1995).

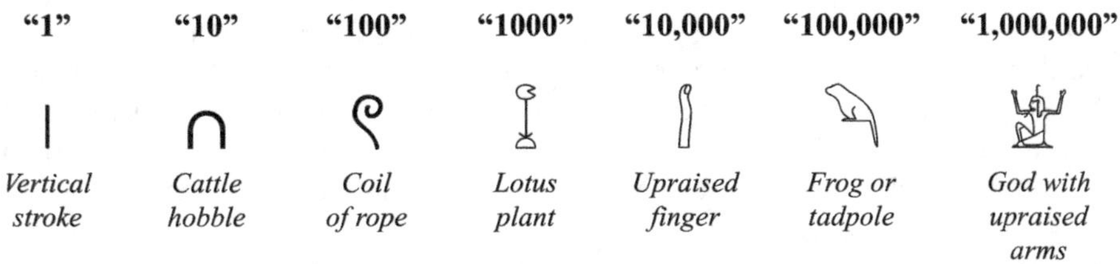

Fig. 7.1 Egyptian hieroglyphs for numbers. Like Aramaic were, these numbers too were ordered from right to left. For example, the number 55,874 is written 𓏤𓏤𓏤𓏤𓎆𓎆𓎆𓎆𓎆𓎆𓎆𓍢𓍢𓍢𓍢𓍢𓍢𓍢𓍢𓆼𓆼𓆼𓆼𓆼𓂭𓂭𓂭𓂭𓂭; in order from largest to smallest (right to left), this is *five* fingers, *five* lotuses, *eight* coils, *seven* hobbles, and *four* strokes. Image by the author. The signs are Unicode characters

Original systems of writing are thought to have emerged in Mesopotamia, Egypt, and China, though granted, the idea of writing could have diffused between all three, given their geographic and temporal proximity (a fourth epicenter, Mesoamerica, is considered truly independent and is discussed separately).[8] All three systems consisted of small pictures and conventions that meant what they looked like, or things related to them. Over time, as proficiency in writing characters developed, writers also gained the ability to tolerate ambiguity in character form.[9] Simply, characters could deform but still be recognized. This meant that characters could be written not just sloppily but also faster; in Egypt, the result was hieratic, a cursive form of hieroglyphs.[10] Characters also eventually became recognized not by what they resembled but through combinations of their structural features and spatial relations. Neurologically speaking, this recognition is a function of the fusiform gyrus, a part of the temporal lobe that recognizes objects and faces; when trained to recognize written characters, it's called the Visual Word Form Area.[11] As characters became recognized through their features and relations, the need for them to resemble the objects they once depicted relaxed; in Egypt, the result was demotic, an abstract descendant of the original set of signs.[12] It's not certain that all writing systems involve both processes, or follow them in the same order; Chinese, for example, doesn't seem to have a cursive form, and in Mesopotamia, cursive developed many centuries after characters lost their depictiveness. The order of these developments in Egyptian writing may reflect its mandate for ideal form, with hieroglyphs acting to anchor the forms of later scripts.

---

[8] Senner (1989).

[9] James and Engelhardt (2012); Longcamp et al. (2005); Mueller and Oppenheimer (2014); Sülzenbrück et al. (2011).

[10] Overmann (2016, 2022).

[11] Coltheart (2014); Dehaene and Cohen (2007); McCandliss et al. (2003).

[12] Overmann (2016, 2022).

Written numbers follow the same general trends. Once written as notations, numerical signs become more ambiguous and then less depictive. However, numerals also differ from other writing because of the involvement of a special perceptual system for quantity. This ability, which humans share with other species, is often called "the number sense,"[13] a name that is misleading because numbers (unlike quantities) don't exist in nature. Because of the so-called number sense, we recognize small quantities (up to *about three or four*) immediately and unambiguously. This is why signs for small numbers typically consist of *one*, *two*, or *three* repeated elements: for example, the vertical wedges of cuneiform (𒁹, 𒈫, and 𒐈) and hieroglyphs (𓏺, 𓏻, and 𓏼) and the horizontal lines of Chinese *Hanzi* (一, 二, and 三). Because small quantities are so recognizable, the forms of these signs have conserved over thousands of years, despite being borrowed and adapted by different languages and cultures. Even the modern Western numerals 2 and 3 are cursive forms of the same simple, repeated strokes: writing two or three lines quickly and without picking the pen up off the paper produces the familiar signs 2 and 3.

Large numbers—defined here as *four* and higher—are also initially written as multiple elements (e.g., *five* and *six* are 𒐊 and 𒐋 in cuneiform and 𓏾 and 𓏿 in hieroglyphs). However, we don't recognize quantities above *about four* like we do small quantities. As a result, multi-element signs are initially divided into subgroups whose quantity is recognizable (generally, by twos and threes). Even with subgroups, repetition quickly becomes unmanageable, so conventional bundles are introduced (e.g., *ten* is 𒌋 in cuneiform and 𓎆 in hieroglyphs). Under conditions of practice, proficiency, and training effects in the fusiform gyrus, signs for large numbers ultimately become more ambiguous and less depictive (Fig. 7.2). This change

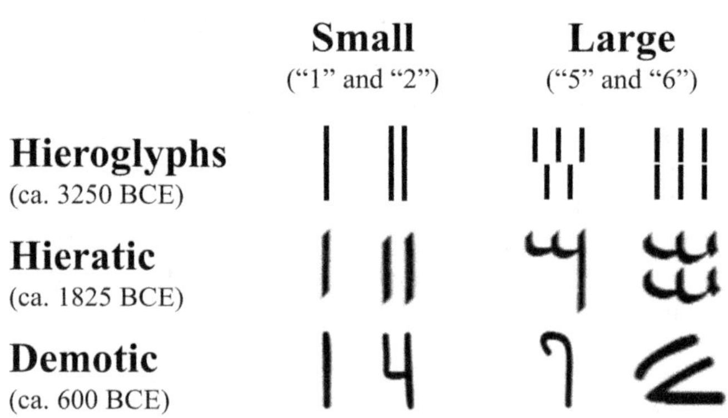

Fig. 7.2 Change in Egyptian numerals over time. Hieroglyphs, the original written form, represented quantity through element repetition (and conventions like those shown in Fig. 7.1). Hieratic was an ambiguous-but-more-quickly-written form of hieroglyphs, while demotic was much less depictive of either precursor script. Signs for small numbers conserve their form to a greater extent than signs for large numbers do. Adapted from Chrisomalis (2010, pp. 35–55)

[13] Dehaene (2011).

avoids the need to count but increases the number of signs that must be learned as conventions.

Signs for numbers differ from writing in other ways. For one thing, once written, numbers don't need to be specified in terms of their sounds values in any language. We saw this with Sumerian number-words, whose sound values weren't recorded until foreign scribes found a reason to do so some centuries later,[14] and Elamite number-words, which apparently no one ever thought to record.[15] For another thing, numbers can be represented without writing at all, as the Sumerians once did with tokens, tallies, and fingers. This means that numbers have a long prehistory before they become written. However, numerical notations are a lot more concise than any of their precursor technologies, and this conciseness enables their accumulation in unparalleled volumes—for example, tables of multiplication and tables of reciprocals accumulate numerical relations at an information density that isn't possible with tokens, let alone tallies or fingers. Scribes created and learned these tables as part of their training.[16] This exposure and knowledge not only caused scribes to think about numbers differently—for example, as related to each other—it also opened up new possibilities for calculating. Notations are fixed, so they aren't suitable for the kind of manual manipulations involved in an abacus or tokens. However, scribes could either look up an answer in a table of relations, or they could provide it from memory. These possibilities influenced calculation toward becoming "knowledge-based," depending to a greater extent on learned relations recalled from memory, compared to the "movement-based" method of the abacus.[17]

The ability to write things down helped to formalize calculation in other ways. Before writing, calculations were simply performed, and final results were kept track of with human memory and objects like tokens and tally marks. Once the ability to write words was available, calculations were recorded, initially as lists of narrative steps akin to a cookbook recipe: "Do such and such to a number and you will get the answer."[18] Over the next several thousand years, these narrative instructions would gradually be refined as equations with operational signs.

References

Chrisomalis, S. (2010). *Numerical notation: A comparative history*. Cambridge University Press.

Coltheart, M. (2014). The neuronal recycling hypothesis for reading and the question of reading universals. *Mind & Language, 29*(3), 255–269.

[14] Damerow (1988); Edzard (1980).

[15] Tavernier (2020).

[16] Postgate (2013); Proust et al. (2007).

[17] Overmann (2019).

[18] Devlin (2003), p. 1.

Damerow, P. (1988). Individual development and cultural evolution of arithmetical thinking. In S. Strauss (Ed.), *Ontogeny, phylogeny, and historical development: The Tel Aviv annual workshop in human development* (Vol. 2, pp. 125–152). Ablex Publishing Corporation.

Dehaene, S. (2011). *The number sense: How the mind creates mathematics* (Revised). Oxford University Press.

Dehaene, S., & Cohen, L. (2007). Cultural recycling of cortical maps. *Neuron, 56*(2), 384–398.

Devlin, K. (2003). *Mathematics: The science of patterns: The search for order in life, mind and the universe*. Henry Holt and Company.

Edzard, D. O. (1980). Sumerisch 1 bis 10 in Ebla [Sumerian 1 to 10 in Ebla]. *Studi Eblaiti, III*(5–8), 121–127.

Gardiner, A. (1950). *Egyptian grammar* (2nd ed.). Oxford University Press.

Godley, A. D. (1922). *Herodotus, The histories*. Harvard University Press.

Hoffmann, F. (2012). Hieratic and demotic literature. In C. Riggs (Ed.), *The Oxford handbook of Roman Egypt* (pp. 543–562). Oxford University Press.

Imhausen, A. (2016). *Mathematics in ancient Egypt: A contextual history*. Princeton University Press.

James, K. H., & Engelhardt, L. (2012). The effects of handwriting experience on functional brain development in pre-literate children. *Trends in Neuroscience and Education, 1*(1), 32–42.

Longcamp, M., Zerbato-Poudou, M.-T., & Velay, J.-L. (2005). The influence of writing practice on letter recognition in preschool children: A comparison between handwriting and typing. *Acta Psychologica, 119*(1), 67–79.

Lopriano, A. (1995). *Ancient Egyptian: A linguistic introduction*. Cambridge University Press.

McCandliss, B. D., Cohen, L., & Dehaene, S. (2003). The visual word form area: Expertise for reading in the fusiform gyrus. *Trends in Cognitive Sciences, 7*(7), 293–299.

Menninger, K. (1992). *Number words and number symbols: A cultural history of numbers* (P. Broneer, Trans.). Dover Publications.

Mueller, P. A., & Oppenheimer, D. M. (2014). The pen is mightier than the keyboard: Advantages of longhand over laptop note taking. *Psychological Science, 25*(6), 1159–1168.

Neugebauer, O. E. (1969). *The exact sciences in antiquity* (2nd ed.). Dover Publications.

Overmann, K. A. (2016). Beyond writing: The development of literacy in the ancient Near East. *Cambridge Archaeological Journal, 26*(2), 285–303.

Overmann, K. A. (2019). *The material origin of numbers: Insights from the archaeology of the ancient Near East*. Gorgias Studies in the Ancient Near East 14. Gorgias Press.

Overmann, K. A. (2022). Early writing: A cognitive archaeological perspective on literacy and numeracy. *Visible Language, 56*(1), 8–44.

Postgate, J. N. (2013). *Bronze Age bureaucracy*. Cambridge University Press.

Proust, C., Donbaz, V., Dönmez, A., & Cavigneaux, A. (2007). *Tablettes mathématiques de Nippur [Mathematical tablets from Nippur]* (Vol. 18). Institut Français d'Études anatoliennes Georges-Dumézil.

Rawlinson, G. (1875). *History of Herodotus* (Vol. II, 3rd ed.). John Murray.

Senner, W. M. (1989). Theories and myths on the origins of writing: A historical overview. In W. M. Senner (Ed.), *The origins of writing* (pp. 1–26). University of Nebraska Press.

Smith, D. E. (1925). *The history of mathematics. Vol.* 2: *Special topics of elementary mathematics*. Ginn and Company.

Sülzenbrück, S., Hegele, M., Rinkenauer, G., & Heuer, H. (2011). The death of handwriting: Secondary effects of frequent computer use on basic motor skills. *Journal of Motor Behavior, 43*(3), 247–251.

Tavernier, J. (2020). Elamite. In R. Hasselbach-Andee (Ed.), *A companion to ancient Near Eastern languages* (pp. 163–184). John Wiley & Sons.

Von Dassow, E. (Ed.). (1998). *The Egyptian* Book of the Dead, *the book of going forth by day, being the Papyrus of Ani (royal scribe of the divine offerings) written and illustrated circa 1250 B.C.E., by scribes and artists unknown* (R. O. Faulkner & O. J. Goelet, Trans., 2nd ed.). Chronicle Books.

Part II
Number Systems in Africa

Chapter 8
The Ishango Bone

Abstract Before the earliest unambiguous numbers in Mesopotamia, there are earlier hints of prehistoric numbers. One is a 25,000-year-old bone from Ishango, a site on a river that flows into Lake Mwitanzige, part of the so-called headwaters of the Nile. This artifact bears three rows of marks that are divided into groups by wider spacing. The groups on the first row suggest the prime numbers between *ten* and *twenty* in ascending order; those of the second row suggest adding and subtracting *one* from *ten* and *twenty*, and those of the third row suggests halving and doubling, but inconsistently. These groupings seem unlikely to have been produced by chance, and scholars have speculated that they show the prehistoric development of complex mathematical concepts. However, as notations, the marks are unbundled and thus cumbersome, a trait that is difficult to reconcile with the notion they represent complex mathematical concepts. The artifact also has a piece of quartz affixed to one end, suggesting it may have been an engraving tool. The marks on a second bone from the same site are not mathematically interesting. This evidence is ultimately inconclusive for prehistoric numbers.

Keywords Prime numbers · Divisibility · Bundling · Restricted counting · Numerical conceptualization

The world's earliest unambiguous numbers, the Mesopotamian impressions, are fairly recent, dated to only some 5000 or 6000 years ago. This late emergence suggests several possibilities: Perhaps Palaeolithic peoples didn't have many numbers, as is true of some societies today in South America and Australia. Or maybe they had numbers but didn't represent them with material forms that would preserve archaeologically, as is the case for contemporary societies that use only their fingers for counting. Or possibly they had numbers and did represent them with durable materials like bone, but the results don't look like numbers to our modern eyes. This is plausible, since it's difficult—if not impossible—for us to know what a group of linear marks once meant to their ancient makers when such marks carry no clues to their meaning. Such ambiguous marks are found all over the world, but no artifacts older than the Mesopotamian impression contains the characteristic groupings that

K. A. Overmann, *Cultural Number Systems*, Interdisciplinary Contributions to Archaeology, https://doi.org/10.1007/978-3-031-83383-0_8

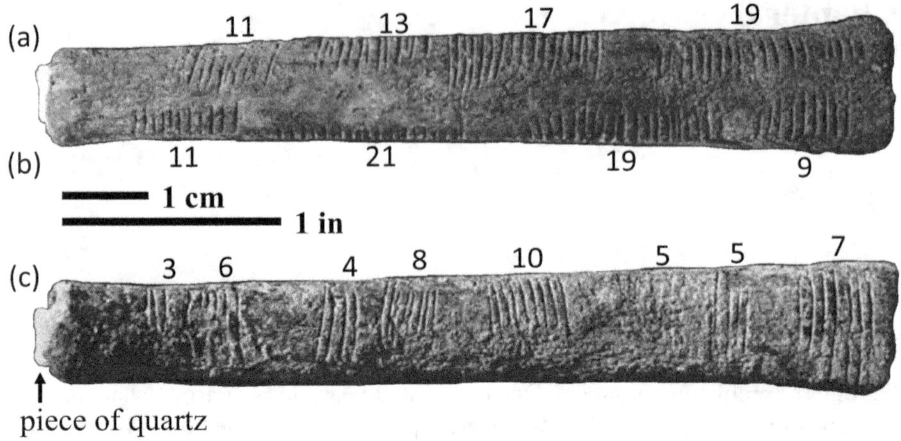

Fig. 8.1 The first Ishango bone. The artifact is just under 4 inches (10 cm) in length and just over half an inch (1.6 cm) wide; the marks range from 0.08 to half an inch (2–13 mm) in length. Adapted and relabeled from de Heinzelin (1962, p. 114, left image)

unambiguously identify its marks as numbers. One of the most tantalizing because it hints at possible numerical grouping is a 25,000-year-old bone from Ishango,[1] a site on a river that flows into Lake Mwitanzige, part of the so-called headwaters of the Nile.

The three rows of marks on the Ishango bone suggest not just quantities but complex mathematical concepts. The incisions are roughly parallel to one another and are periodically separated into groups by wider spacing. As was initially noted by the bone's discoverer, the Belgian geologist Jean de Heinzelin, as well as many scholars since, the first row (Fig. 8.1a) has four groups with 11, 13, 17, and 19 marks; these are the prime numbers between *ten* and *twenty* in ascending order. The second row (Fig. 8.1b) has four groups of 11, 21, 19, and 9 marks; these are the numbers produced by adding and subtracting *one* from *ten* and *twenty*. The third row (Fig. 8.1c) contains eight groups whose quantities—3, 6, 4, 8, 10, 5, 5, and 7—suggest halving and doubling, though inconsistently.[2]

The seeming impossibility of producing such sequences by chance has fueled intense speculation about mathematics having deep African roots, particularly for Egypt, given the riverine connection. De Heinzelin himself believed that the artifact might have been a tool for engraving or tattooing because of the piece of quartz affixed to one end, with the marks possibly representing decimal numbers and the concept of prime numbers.[3] Alexander Marshack, the American archaeologist known for his work on prehistoric notational systems, thought that the marks recorded a six-month lunar calendar, though he based his interpretation in part on

[1] Brooks and Smith (1987).

[2] De Heinzelin (1962).

[3] De Heinzelin (1962).

marks that are either too small or too shallow to be seen except by microscope.[4] Claudia Zaslavsky, an American ethnomathematician known for her work on African mathematics, asserted that the bone was most likely made by a woman to keep track of her menstrual cycles.[5] Within the past decade, other researchers have claimed that the bone has duodecimal (base-12) numbers with sub-bases of 3 and 4, and that their layout enables multiplication in a manner analogous to a slide rule.[6] Most recently, the bone has been proposed to comprise a mathematical sieve for calculating prime numbers, and that possibly when it's conjoined with a much-less-famous and still-unpublished second bone found in the same archaeological layer,[7] it acts as a prime sieve with a formula structured on a double helix like the one found in DNA.[8]

Not everyone agrees.[9] As one historian of mathematics succinctly put it, "A single bone with suggestive markings raises interesting possibilities of a highly developed sense of arithmetical awareness; it does not provide conclusive evidence."[10]

There are good reasons to be skeptical about the meaning of the bone's marks. Bones are inscribed for all sorts of purposes that aren't necessarily numerical: mnemonic prompts, calendrical recording, wayfinding points, cultural designs affording safe passage to messengers, music-making (the instrument known as a rasp), and grip improvement (achieved by roughening the surface) of tools like awls.[11] As was noted, the bone has a piece of quartz affixed to one of its ends, and this feature seems much more consistent with a purpose like engraving than it does counting—and hence, the idea that marks might represent a utilitarian purpose like grip improvement is not entirely irrelevant.

The marks on the second bone aren't nearly as interesting, numerically speaking, and this raises the possibility that the marks on the first bone are just statistical anomalies, albeit highly improbable ones. The museum holding both artifacts, the Royal Belgian Institute of Natural Sciences, has stated only that the second bone has 90 marks that differ in length sufficiently to characterize them as "major" and "minor," and further noted that de Heinzelin studied but never published it, nor had anyone else as of 2018.[12] The artifact has been described as containing six rows with 20, 6, 18, 6, 31, and 8 long and short marks spaced irregularly (Fig. 8.2). The diagram apparently omits a mark from Row 1,[13] which brings the total to 89 when

[4] Marshack (1991).

[5] Zaslavsky (1992, 1999).

[6] ADIA-RBINS (2015), p. 20); also see Pletser and Huylebrouck (1999, 2015).

[7] ADIA-RBINS (2018); as of 2023, the page on the second Ishango bone is no longer available on the museum's website.

[8] Born (2019); Kamalu (2021).

[9] E.g., Keller (2010).

[10] Joseph (2011), p. 33.

[11] Hugh-Jones (2016); Kelly (2020); Marshack (1985); Reese (2002); Tixier (1974).

[12] ADIA-RBINS (2018).

[13] Born (2019).

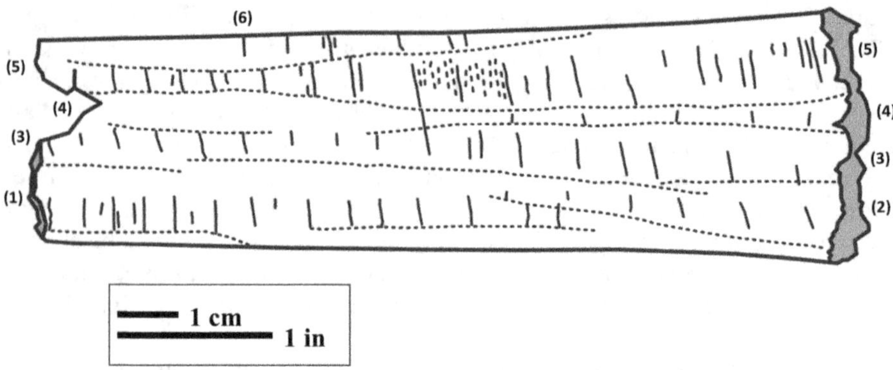

Fig. 8.2 The second Ishango bone, visualized as an unrolled 3D drawing. The artifact is 5.5 inches (14 cm) in length (Born, 2019); the marks range in size from 0.30 to 0.06 inches (7.6 to 1.5 mm). Copyright RBINS (see ADIA-RBINS, 2018) and used by permission

added together; de Heinzelin possibly counted one of the several shallow incisions marked with dashed lines in the middle of Row 5 to reach his total of 90. Of note, these totals require counting separately any line that crosses multiple rows (marked with arrows in the figure), as well as two facing marks in Row 5 that might make a single line if they were connected.

While the quantities of marks on the famous first bone are indeed mathematically suggestive, only one group exceeds the upper limit of *twenty*, and then, by only *one*. This makes the marks generally consistent with what is called restricted counting. A *restricted counting system* is one in which numbers count no higher than *twenty*,[14] though such systems may count no higher than *two* or *three*, consistent with the limits on perceptible quantity. Today, we associate restricted counting with societies that are small and relatively isolated, conditions under which numbers and counting are neither needed nor motivated, let alone complex concepts like subtraction, halving and doubling, and indivisibility. Certainly, the site's archaeological reports show a small society with a semisedentary lifestyle at the river's edge,[15] which would generally be consistent with restricted counting. In comparison, complex numbers develop once societies become large, interconnected, and complex, and need numbers and counting to manage that complexity.[16] Conditions of complexity were certainly not true of prehistoric Ishango. The mystery runs deeper still. Lines incised in bone can't be moved or removed, so the device itself could not have been used for calculating. Complex concepts like divisibility require a technology that can be manipulated, like the tokens used in Mesopotamia[17] or the pebbles said by Herodotus

[14] Comrie (2011, 2013).

[15] Brooks et al. (2006); Brooks and Smith (1987).

[16] Overmann (2013).

[17] Høyrup (2002); Woods (2017).

to have been used for calculating in Egypt.[18] If such means were available, we'd again expect the artifact to be associated with a complex society, but it simply wasn't.

Given the complexity of the purported concepts, we might also expect the marks to be bundled, perhaps with a longer mark that meant *five* or *ten* of the shorter marks. Bundling makes numerical information more concise, easier to move when calculating, and more accessible when being written or read. Since bundling wasn't used, the groups of marks are relatively clumsy indicators of particular quantities. The intriguing combination of conceptual complexity and notational clumsiness is difficult to explain. Possibly the lack of bundling isn't as great an impediment to mathematical conceptualization as we might think. Perhaps the mathematical concepts were worked out with a manipulable technology and later recorded using an uncouth notation; given frequency of use and enough time, such notational clumsiness would eventually motivate the development of conventions for bundled values. However, there's no evidence that such a system of notations—or mathematics—ever developed in the vicinity of Ishango. We are left to marvel at a true prehistoric mystery.

References

ADIA-RBINS. (2015). *Discover Ishango*. Association for the Dissemination of Archaeological Information (ADIA) and Royal Belgian Institute of Natural Sciences (RBINS). https://www.naturalsciences.be/sites/default/files/DiscoverIshango.pdf

ADIA-RBINS. (2018). *The second Ishango bone*. Association for the Dissemination of Archaeological Information (ADIA) and Royal Belgian Institute of Natural Sciences (RBINS). https://ishango.naturalsciences.be/en/en-ishango-21.html

Born, C. (2019). *The secret of Ishango bones: On the helix structure of prime numbers, a mathematical essay* (Preprint). https://doi.org/10.31730/osf.io/yz6nq

Brooks, A. S., & Smith, C. C. (1987). Ishango revisited: New age determinations and cultural interpretations. *African Archaeological Review, 5*, 65–78.

Brooks, A. S., Nevell, L., Yellen, J. E., & Hartman, G. (2006). Projectile technologies of the African MSA: Implications for modern human origins. In E. Hovers & S. L. Kuhn (Eds.), *Transitions before the transition: Evolution and stability in the Middle Palaeolithic and Middle Stone Age* (pp. 233–255). Springer.

Comrie, B. (2011). *Typology of numeral systems*. Trends in Linguistics. Studies and Monographs 118. Max Planck Institute. https://lingweb.eva.mpg.de/channumerals/TypNum_Latest_21ho.pdf

Comrie, B. (2013). Numeral bases. In M. S. Dryer & M. Haspelmath (Eds.), *World atlas of language structures online*. Max Planck Institute for Evolutionary Anthropology. http://wals.info/chapter/131

De Heinzelin, J. (1962). Ishango. *Scientific American, 206*(6), 105–116.

Høyrup, J. (2002). *Lengths, widths, surfaces: A portrait of Old Babylonian algebra and its kin*. Springer.

Hugh-Jones, S. (2016). Writing on stone; writing on paper: Myth, history and memory in NW Amazonia. *History and Anthropology, 27*(2), 154–182.

Joseph, G. G. (2011). *The crest of the peacock: Non-European roots of mathematics* (3rd ed.). Princeton University Press.

[18] Smith (1925).

Kamalu, C. (2021). *The Ishango bone: The world's first known mathematical sieve and table of the small prime numbers* (Preprint). https://osf.io/preprints/africarxiv/6z2yr/download

Keller, O. (2010). *Les fables d'Ishango, ou l'irrésistible tentation de la mathématique-fiction [Ishango's fables, or the irresistible temptation of mathematical fiction]*. Bibnum. http://journals.openedition.org/bibnum/889

Kelly, P. (2020). Australian message sticks: Old questions, new directions. *Journal of Material Culture, 25*(2), 133–152.

Marshack, A. (1985). A lunar-solar year calendar stick from North America. *American Antiquity, 50*(1), 27–51.

Marshack, A. (1991). *The roots of civilization: The cognitive beginnings of man's first art, symbol and notation* (Revised). Moyer Bell.

Overmann, K. A. (2013). Material scaffolds in numbers and time. *Cambridge Archaeological Journal, 23*(1), 19–39.

Pletser, V., & Huylebrouck, D. (1999). The Ishango artefact: The missing base 12 link. *Forma, 14*, 339–346.

Pletser, V., & Huylebrouck, D. (2015). *Contradictions et étroitesse de vues dans "Fables d'Ishango, ou l'irrésistible tentation de la mathématique-fiction," réponses et mises au point [Contradictions and narrow-mindedness in "Ishango's Fables, or the irresistible temptation of mathematical fiction," answers and clarifications]*. https://arxiv.org/ftp/arxiv/papers/1607/1607.00860.pdf

Reese, D. S. (2002). On the incised cattle scapulae from the East Mediterranean and Near East. *Bonner Zoologische Beitrage, 50*, 183–198.

Smith, D. E. (1925). *The history of mathematics. Vol.* 2: *Special topics of elementary mathematics*. Ginn and Company.

Tixier, J. (1974). Poinçon décoré du Paléolithique Supérieur à Ksar'Aqil (Liban) [Decorated punch from the Upper Paleolithic at Ksar'Aqil (Lebanon)]. *Paléorient, 2*(1), 187–192.

Woods, C. (2017). The abacus in Mesopotamia: Considerations from a comparative perspective. In L. Feliu, F. Karahashi, & G. Rubio (Eds.), *The first ninety years: A Sumerian celebration in honor of Miguel Civil* (pp. 416–478). Walter de Gruyter.

Zaslavsky, C. (1992). Women as the first mathematicians. *International Study Groups on Ethnomathematics Newsletter, 7*(1). http://web.nmsu.edu/~pscott/isgem71.htm

Zaslavsky, C. (1999). *Africa counts: Number and pattern in African cultures*. Lawrence Hill Books.

Chapter 9
Yoruba Numbers

Abstract The Yoruba of West Africa have a number system based on counting cowries, shells used as currency for millennia. Up to the number 400, Yoruba numbers are organized by twenties, an organization typically based on counting with the fingers and toes. Starting with 200, Yoruba numbers are organized by tens, suggesting a historically recent expansion based on counting bundles of cowrie shells in conjunction with selling slaves to European traders. Yoruba numbers involve more than a typical amount of subtractive formations, an operation known as back-counting. They are also curiously flexible in how they can be formed, permitting any particular number to be expressed in an unusual number of alternative but well-formed ways. A typical example is 19,669, which can be expressed in 11 different ways. This flexibility suggests that Yoruba numbers are conceived of as arrays of combinatorial possibilities, rather than as points along a linear continuum as in the Western number line. An array of cowrie shells has also been described as being used like an abacus, and this visual arrangement may help influence these numbers to be conceptualized as arrays.

Keywords Cowrie counting · Vigesimal · Back-counting · Combinatorics · Array

Across the continent and to the northwest of Ishango lies West Africa, the coastal area where Europe bought slaves for its New World colonies from the seventeenth to the nineteenth centuries CE. While many types of goods were used to purchase human beings, the preferred currency was the cowrie shell, which European ships brought in by the ton.[1] Cowries are sea mollusks that, depending on the species, range in size from half an inch to several inches. They were used as currency in West Africa for centuries: Ibn Battuta, the medieval scholar who traveled the world between 1325 and 1354, visited West Africa in 1351 and noted that cowries were used as currency in Mali, particularly in the city of Gao.[2] The use of cowries has had interesting effects on West African number systems. An example is the traditional number system of the Yoruba, a people whose homeland spans Nigeria, Benin, and

[1] Hogendorn and Johnson (1986).

[2] Battuta (1994), p. 827.

K. A. Overmann, *Cultural Number Systems*, Interdisciplinary Contributions to Archaeology, https://doi.org/10.1007/978-3-031-83383-0_9

Togo. Their traditional number system has been called "one of the most peculiar in existence,"[3] perhaps because numbers are envisioned as arrays of combinatorial possibilities, rather than as linear progressions.

Yoruba is often classified as a vigesimal (base-20) system, and indeed, it may have begun as a digit-tally system of counting on the fingers and toes. Numbers up to *twenty squared* (*irínwó*) are formed on multiples of *twenty*. However, overlaying this structure is another sequence of numbers that starts at *two hundred* and increases by multiples of tens, consistent with the decimal numbers used by European traders (Fig. 9.1). *Five* strings of *forty* cowries made a smaller bundle of *two hundred* cowries (*igba*); *ten* of these smaller bundles made a larger bundle of *two thousand* cowries (*ẹgbẹ̀wa*); and *ten* of the larger bundles made a bag of *twenty thousand* cowries (*ọ̀kẹ́*), "a reasonable load for a man [to carry] over long distances" at 50 to 100 pounds (23–46 kg).[4] When Glélé, the king of neighboring Dahomey, attacked the Yoruba in 1864,[5] he reportedly lost men numbering "two heads [*ẹgbẹ̀wa*, worth 2000 cowries each], twenty strings [worth 40 cowries each], and twenty [single] cowries," for a total of 4820 men.[6] Confusion over the way that cowrie bundles represent numbers is shown by this total elsewhere being transcribed as 6820 or

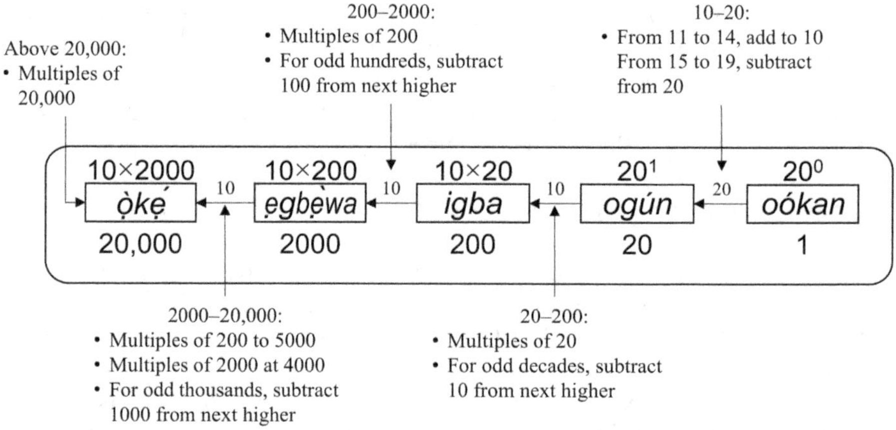

Fig. 9.1 Yoruba numbers and the cowrie bundles they are based on. Image by the author based on data from multiple sources. A version was previously published in Overmann (2023, Fig. 12.7, p. 299)

[3] Conant (1896), p. 70.

[4] Hogendorn and Johnson (1986), p. 76; also see Mann (1886); Yang (2019).

[5] Burton (1864).

[6] Conant (1896), pp. 70–71.

6821.[7] Here a Yoruba proverb might apply: "You cannot multiply *nine* by *nine*," which essentially means, "You are a dunce."[8]

Yoruba numbers involve a lot of subtraction. Subtraction in and of itself isn't unusual, as it's found in Latin, where the words for *eighteen* and *nineteen*, *duodeviginti* and *undeviginti*, mean *two from twenty* and *one from twenty*. Sumerian also had subtractive forms, as 18 and 19 might be formed as *20 minus 2* and *20 minus 1* in mathematical texts.[9] Yet Yoruba is more subtractive than most. *Fifteen* through *nineteen* are formed by subtracting the appropriate small number from *twenty*: *Fifteen* is *five from twenty*, *sixteen* is *four from twenty*, and so on. The odd decades, hundreds, and thousands are also formed by subtracting, respectively, *ten*, *hundred*, and *thousand* from the next higher even amount: *Seventy* is *ten from eighty*, *seven hundred* is *one hundred from eight hundred*, *seven thousand* is *one thousand from eight thousand*, and so on.[10] These formations are consistent with the traditional method of cowrie counting, where *five* cowries were taken up in each hand to form groups of *ten*, and groups of *ten* were used to form the bundles of *twenty*, *two hundred*, *two thousand*, and *twenty thousand* shown in Fig. 9.1.[11] For interim amounts, it's sometimes easier to remove cowries from the next higher group than to add more to the next lower, with the point where the difficulty becomes about equal being somewhere in the middle.[12]

Yoruba is curiously flexible in the way it forms its higher numbers. In Yoruba, it's possible to express the same number in a well-formed fashion in an unusual number of ways, combinatorics involving the multiplication, addition, and subtraction of different cowrie bundles. For example, there are at least 11 different ways to express 19,669 (Table 9.1), a complex number that we would typically express as *nineteen thousand, six hundred and sixty-nine*. For simpler numbers like 1,200,000, we might exchange the decimal powers back and forth to create alternative expressions like *one point two million*, *one million two hundred thousand*, or *twelve hundred thousand*.[13] However, we wouldn't say *two six hundred thousands* or *one eight hundred thousand from two million*. Yoruba's ability to combine and sequence arithmetical operations to construct complex numerical expressions is unparalleled among the world's number systems.

How quickly an average Yoruba individual might be able to express multiple well-formed versions of a complex number like 19,669 is a very good question. The eleven alternatives in Table 9.1 were split between two articles published nearly 40 years apart, which suggests that some of the names might be difficult to generate, even if all of them are ultimately well formed. Unfortunately, the rapid and ongoing

[7] Burton (1864), p. 316, (1924), p. 117.

[8] Burton (1924), p. 117; Raum (1938).

[9] Friberg (2009); Friberg and Al-Rawi (2016); also see Menninger (1992).

[10] Armstrong (1962); Crowther (1852); Ẹkundayọ (1977).

[11] Mann (1887), p. 62.

[12] Mann (1887), p. 62.

[13] Chrisomalis (2020).

Table 9.1 Eleven ways to form 19,669 in Yoruba (Akinadé & Ọdẹ́jọbí, 2014; Ẹkundayọ, 1977; also see Burton, 1863; Crowther, 1852)

Yoruba expression for 19,669	Numerical meaning
Ọkẹ́ kan ó dín ọ̀ọ́dúnrun ó dín ọ̀kànlélọ́gbọ̀n	(20,000 × 1) – 300 – (1 + 30)
Ọkẹ́ kan ó dín irinwó ó lé ọkaàn dínláàdọ́rín	(20,000 × 1) – 400 + (1 from (10 from (20 × 4)))
Ọkẹ́ kan ó dín ojìlélọdúnrún ó lé mẹsán	(20,000 × 1) – ((20 × 2) + 300) + 9
Ọkẹ́ kan ó dín ọ̀tàdínirinwó ó lé mẹ́sàn	(20,000 × 1) – (60 from 400) + 9
Ẹ̀ẹ́dẹ́gbàáwàá ó lé ójìlélẹgbẹta ó lé mọkandínlọgbọn	(1000 from (2000 × 10)) + ((20 × 2) + (200 × 3)) + (1 from 30)
Ẹ̀ẹ́dẹ́gbàáwàá ó lé ọ̀tàlélẹgbẹta ó lé mẹ́sàn	(1000 from (2000 × 10)) + ((20 × 3) + (200 × 3)) + 9
Ẹ̀ẹ́dẹ́gbàáwàá ó lé ọ̀rinlélẹgbẹ̀ta ó dín ókànlàá	(1000 from (2000 × 10)) + ((20 × 4) + (200 × 3)) – (1 + 10)
Ẹ̀ẹ́dẹ́gbàáwàá ó lé ẹgbẹ̀ta ó lé ọ̀kàndínláádọrin	(1000 from (2000 × 10)) + (100 from (200 × 4)) – (1 + 30)
Ẹ̀ẹ́dẹ́gbàáwàá ó lé ẹ̀ẹ́dẹ̀gbẹrin ó dín ọ̀kànlélọ́gbọ̀n	(1000 from (2000 × 10)) + (200 × 3) + (1 from (10 from (20 × 4)))
Ẹ̀ẹ́dẹ́gbọkandínlọgọrún ó dín mọkanlélọgbọn	(100 from (200 × (1 from (20 × 5)))) – (1 + 30)
Ẹgbèjìdínlọgọrún ó lé mọkandínlaadọrin	(200 × (2 from (20 × 5))) + (1 from (10 from (20 × 4)))

displacement of the traditional Yoruba system by Western numbers, along with various initiatives to revise Yoruba numbers as a standard decimal system,[14] mean the unique ability to form numbers with so many different combinations will most likely be lost.

The Yoruba made a type of abacus out of cowries. The Belgian mathematician Dirk Huylebrouck described this abacus as a grid composed of *twenty* groups of twenty cowrie shells arranged in a rectangle (Fig. 9.2).[15] Intriguingly, multiplication on this grid produces answers consistent with Yoruba spoken numbers and their combinatorial logic.

In actuality, any number divisible by *four* could act as the base of such a grid, *four* because it permits the shells to be arranged as a rectangle. The grid should also be big enough in terms of the number of its elements to accommodate large numbers, but not so big that the amount of shells becomes unmanageable (400 arguably tests this limit). Both multiplicand and multiplier should also be fairly large (i.e., close to the selected base, like 19 and 17 are to 20), because this minimizes the number of shells that need to be moved. Multiplying even 15 × 15 requires moving nearly half the shells.

[14] Adeyinka (2010).

[15] Huylebrouck (2006, 2019).

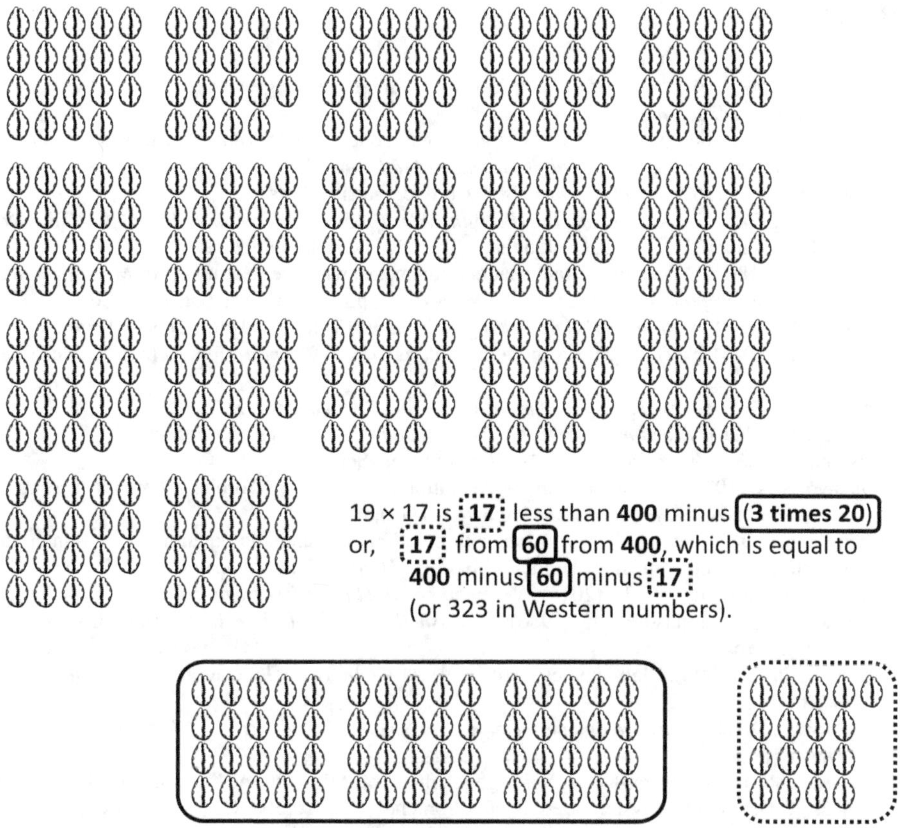

Fig. 9.2 Cowrie grid showing the final result of multiplying 19 × 17. Image by the author based on material in Huylebrouck (2006, Fig. 22 right, p. 156, and 2019, Figs. 6.5, 6.6, and 6.7, pp. 111–113). A version was previously published in Overmann (2023, Fig. 12.8, p. 305)

Huylebrouck unfortunately provided no description of where he encountered the cowrie abacus, since it was not mentioned in the 1887 paper by the missionary Adolphus Mann that he cited as his source.[16] Nor does he say how old the practice might have been; how widespread it actually was; whether it included algorithms for operations like division, addition, and subtraction; or how and why it was employed, apart from the single example of multiplying 19 × 17. Perhaps there were ways of working around the limitations of size, movements, and use noted above. However, as no other author has described the grid in such detail, we are left to contemplate another African numerical mystery.

[16] Mann (1887).

References

Adeyinka, A. A. (2010). Proposal for a Yoruba decimal counting system. *Ghana Journal of Education and Teaching, 11*, 221–235.

Akinadé, O. O., & Ọdẹ́jọbí, A. (2014). Computational modelling of Yorùbá numerals in a number-to-text conversion system. *Journal of Language Modelling, 2*(1), 167–211.

Armstrong, R. G. (1962). *Yoruba numerals*. Oxford University Press.

Battuta, I. (1994). *The travels of Ibn Battuta, AD 1325–1354* (Vol. IV, H. A. R. Gibb & C. F. Beckingham, Eds., C. F. Beckingham, Trans.). The Hakluyt Society.

Burton, R. F. (1863). Notes on certain matters connected with the Dahoman. In *Memoirs read before the Anthropological Society of London* (Vol. 1, pp. 308–321). Trübner and Co.

Burton, R. F. (1864). *A mission to Gelele, King of Dahome* (Vol. II, 2nd ed.). Tinsley Brothers.

Burton, R. F. (1924). *Selected papers on anthropology, travel & exploration*. Robert M. McBride and Company.

Chrisomalis, S. (2020). *Reckonings: Numerals, cognition, and history*. MIT Press.

Conant, L. L. (1896). *The number concept: Its origin and development*. Macmillan.

Crowther, S. (1852). *A grammar of the Yoruba language*. Seeleys.

Ẹkundayọ, S. A. (1977). Vigesimal numeral derivational morphology: Yoruba grammatical competence epitomized. *Anthropological Linguistics, 19*(9), 436–453.

Friberg, J. (2009). A geometric algorithm with solutions to quadratic equations in a Sumerian juridical document from Ur III Umma. *Cuneiform Digital Library Journal, 3*, 1–27.

Friberg, J., & Al-Rawi, F. N. H. (2016). *New mathematical cuneiform texts*. Springer.

Hogendorn, J., & Johnson, M. (1986). *The shell money of the slave trade*. Cambridge University Press.

Huylebrouck, D. (2006). Mathematics in (central) Africa before colonization. *Anthropologica et Præhistorica, 117*, 135–162.

Huylebrouck, D. (2019). *Africa and mathematics: From colonial findings back to the Ishango rods*. Springer Nature.

Mann, A. (1886). Die Yoruba-Küste. Vortrag, gehalten von Missionar A. Mann am 18. Dezember 1885. I. Das Zahlen-System der Yoruba-Sprache [The Yoruba Coast. Lecture given by Missionary A. Mann on December 18, 1885. I. The number system of the Yoruba language]. In *Separat-Abdruck aus dem III.-IV. Jahresbericht des Württembergischen Vereins fur Handelsgeographie [Separate reprint from the III.-IV. Annual Report of the Württemberg Association for Commercial Geography]* (pp. 66–73). Württembergischen Vereins für Handelsgeographie.

Mann, A. (1887). Notes on the numeral system of the Yoruba nation. *Journal of the Anthropological Institute of Great Britain and Ireland, 16*, 59–64.

Menninger, K. (1992). *Number words and number symbols: A cultural history of numbers* (P. Broneer, Trans.). Dover Publications.

Raum, O. F. (1938). *Arithmetic in Africa*. Evans Brothers, Ltd.

Yang, B. (2019). *Cowrie shells and cowrie money: A global history*. Routledge.

Chapter 10
Igbo Numbers

Abstract The Igbo are a Nigerian ethnic group whose homeland is south of Yorubaland in West Africa. Like the traditional Yoruba numbers, the traditional Igbo numbers are vigesimal and based on counting cowrie shells. Unlike Yoruba numbers, Igbo numbers use bundles of *six* as the unit of counting with cowries. This collective counting created a second counting sequence: in addition to a sequence for counting things one by one, a second sequence counted units of *six* one by one, and the resultant values were understood as multiplied sixfold (that is, three cowrie bundles would be understood as having the value *eighteen*). Like Yoruba numbers, Igbo numbers can be formed in an unusual number of alternative but well-formed ways. This flexibility does not represent a lack of *lexicalization*, the process whereby names for numbers become standardized in speech. Rather, like the Yoruba people appear to have done with their traditional numbers, the Igbo people may conceive of their traditional numbers as arrays of combinatorial possibilities. This conceptualization suggests an influence of the material form used to represent numbers, the bundles of cowrie shells and the collective method of counting them.

Keywords Cowrie counting · Vigesimal · Collective counting · Combinatorics · Array

Just south of Yorubaland live the Igbo (or Ibo), a Nigerian ethnic group whose traditional numbers are vigesimal, like those in the related Yoruba language (both languages are members of the Volta–Niger family).[1] Like the Yoruba, the Igbo counted cowries, but they did so by grouping them into a new plural unit, *ego*, a bundle of *six nkpulu* or single shells.[2] *Ten ego* made *ego neli* (*ten* sixes or 60 shells); *twenty ego neli* made *ozu ughu* (*twenty* sixties or 1200 shells); twenty *ozu ughu* made *ozu afia* (*twenty* 1200s or 24,000 shells); *twenty ozu afia* made *nnù afia nabo* (*twenty*

[1] Chan (2021).

[2] Basden (1921); Jeffreys (1938, 1948).

K. A. Overmann, *Cultural Number Systems*, Interdisciplinary Contributions to Archaeology, https://doi.org/10.1007/978-3-031-83383-0_10

24,000s or 480,000 shells); and *twenty nnù afia nabo* made *nnù afia nnù* (*twenty* 480,000s or 9,600,000 shells).[3]

The Igbo method of counting was straightforward. Counting began by forming groups of *six* shells (*ego*); these were combined *ten* at a time to form each *ego neli*;[4] these in turn were bundled *twenty* at a time to form each *ozu ughu*, then *twenty* at a time to form each *ozu afia*, and so on. The term *nnù*, which meant 400 (*twenty* squared) in the vigesimal system, was revalued as 200 in the cowrie system. This change yielded cowrie values linked to the original vigesimal numbers by factors of *ten*, *three*, and *two* up to the sixth power of *twenty* (Fig. 10.1). The elegant and harmonious regularity of these relations would undoubtedly have been highly advantageous in moving between the two systems, especially when calculating mentally.

Early observers of the Igbo number systems found them hard to understand. The difficulty was twofold. First, there was the plural unit—*ego*, the group of six—used in counting cowries. Westerners have found the Polynesian decimal system impenetrable for the same reason, and with a similar result: Just as the Polynesian practice of counting pairs with decimal numbers has been mistaken for a vigesimal system (something that occasionally still happens today),[5] the Igbo system of counting sixes with vigesimal numbers was mischaracterized as duodecimal (base-12) or sexagesimal (base-60).[6] In Polynesia, the method of counting didn't merely confuse Westerners about whether the number system was decimal or vigesimal, it also sparked a fantastic story about the New Zealand Māori counting things by elevens

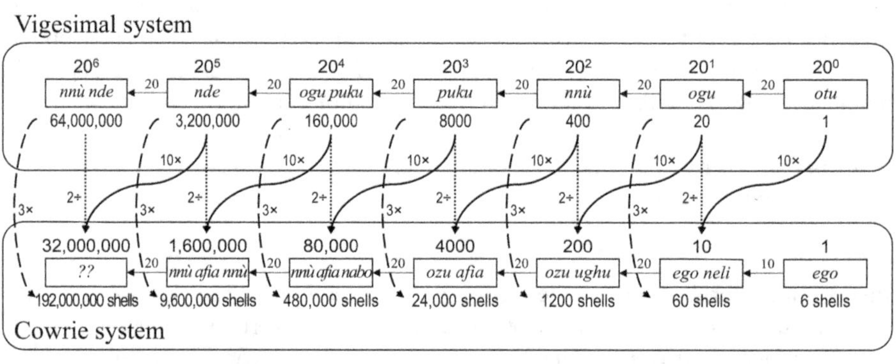

Fig. 10.1 The relations between the Igbo vigesimal and cowrie systems. (Top) The vigesimal exponents (powers of *twenty*), their Igbo names, and the associated decimal values (e.g., 20^2 is 400 or *nnù*). (Bottom) The cowrie bundles, their Igbo names, and the total number of shells in each (e.g., 200 *ego* or *ozu ughu* is 1200 shells). Only the relation between the first vigesimal exponent (20^0, the unit or 1) and *ego* in the cowrie system (the plural unit of *six* shells) is irregular. Thereafter, the relations are 10× (solid line), 3× (dashed line), and 2÷ (dotted line). Image by the author

[3] Jeffreys (1938), p. 224.

[4] Basden (1921), p. 198.

[5] Best (1906); Kendall (1817); Lee (1820); Lemaitre (2004); Nishimoto (2015).

[6] Jeffreys (1948).

(Chap. 31).[7] Similarly, Western observers weren't quite sure what they were seeing with regards to the Igbo number systems. The British colonial administrator and anthropologist Mervyn Jeffreys, for example, knew about the plural unit and saw that it organized the cowrie system by sixes, but concluded that the system was organized by *twelve* or *sixty*, rather than *ten*.[8] Admittedly, Igbo numbers aren't a senary (base-6) system either. It's just difficult to classify number systems that count with plural units, as these fall outside the paradigm wherein objects are counted *one* by *one*. However, following the logic that classifies Polynesian numbers as decimal—even though they are used to count pairs and groups of *four* or *eight* and not just singles—Igbo numbers can be classified as vigesimal while counting with both singles and groups of *six*.

Also contributing to the confusion was the fact that Igbo speakers, like the neighboring Yoruba, varied in the way they formed their numbers:

> Some difficulty is often experienced in ascertaining native methods of counting. For it may be found that there is no set nomenclature, and that though no mistakes are made in actual counting yet the native wording may vary widely in different individuals.
>
> This difficulty can be demonstrated in English. Thus, the term 600 has a definite name in English but a native may express the same total as 50 in twelve places or as 60 in ten places: or as 100 in six places, or as 20 in thirty places. He will state that all are understood and no one method is more accurate or better than another.
>
> (Jeffreys, 1948, pp. 224–225)

Naming variability is the key to understanding Igbo numbers. Jeffreys seems to have interpreted the variability in naming as the absence of settled names. This is characteristic of number systems that are relatively young and have restricted counting ranges. In these systems, it does indeed take a while for names to conventionalize, and then for lexicalization to occur, *lexicalization* being the ability to generate new number-names (like *one thousand, six hundred, and sixty-one*) from memorized words for small and productive numbers (in this example, *one*, *six*, *hundred*, and *thousand*).[9] While there's no way to know how long conventionalization and lexicalization might take—hundreds or perhaps even thousands of years—they likely take less time when numbers are used often, something that is governed by social needs and priorities. Nonetheless, what we see in the Igbo and Yoruba number systems is a vastly different phenomenon: the ability to express the same numbers—and large, complex numbers, at that—in multiple ways, all of them well formed. In other words, it's not that these number systems lack conventions or lexicalization. Rather, they seem to envision numbers as *arrays*, arrangements of objects that represent combinatorial possibilities.

Understanding how this works starts with looking at the way that Western numbers are represented—by means of symbolic numerals like "7"—and then considering what such symbols tells us. In actuality, such symbols don't tell us much because their meaning depends on the user mastering a significant implicit component,

[7] Overmann (2020).

[8] Jeffreys (1948).

[9] Greenberg (1978).

information that isn't directly encoded in the symbol forms themselves. For example, we learn that "7" is the quantity of sets with *seven* members (its *cardinality* or "how many" somethings there are) and the number that falls between "6" and "8" (its *ordinality* or position in a numerical sequence). We learn that "7" is "2" more than "5" and that twice "7" is "14" (its relations with other numbers). And we learn that "7" is valued as "7" in the units position but "70" in the tens position and "$^{7}/_{10}$" when preceded by a decimal point (its place value). But nothing in the symbol form itself tells us any of this information; rather, these are all things that we must know in order to use the symbolic form.

The Greek mathematicians who investigated the properties of numbers more than two thousand years ago understood this. Accordingly, in order to visualize the internal properties of numbers, they represented them with dots.[10] The number *four*, for example, was a square with *two* dots per side (this is the etymology of the term *square* for multiplying a number by itself). The dots revealed internal relations in a way that symbols like "4" (and its equivalent in the symbols used by the ancient Greeks) cannot. Similarly, when the Igbo or Yoruba use cowrie piles and bundles as their material form of representing numbers (or the Yoruba cowrie grid, for that matter), they don't just see the internal relations of each pile or bundle; they also see the potential relations between piles and bundles—and hence, the flexible combinatorics in how large numbers might be put together from those piles and bundles.

Igbo and Yoruba numbers illustrate an important issue. The material forms we use for representing numbers are critical to how we acquire numbers and what we think numbers are. We understand this when it comes to numerals,[11] the form most often used in the Western tradition to represent numbers, but we are less likely to see that the same thing is true for non-notational forms like cowrie bundles. This phenomenon too is related to the material form we use: Since we closely associate numbers with the symbolic notations we use, we tend to be blind to how non-notational forms work in representing numbers.

References

Basden, G. T. (1921). *Among the Ibos of Nigeria: An account of the curious & interesting habits, customs & beliefs of a little known African people by one who has for many years lived amongst them on close & intimate terms*. J. B. Lippincott Company.

Best, E. (1906). Māori numeration: Some account of the single, binary, and semi-vigesimal systems of numeration formerly employed by the Māori. *Transactions and Proceedings of the New Zealand Institute*, *39*, 150–180.

Chan, E. (Ed.). (2021). *Numeral systems of the world's languages*. Max Planck Institute. https://lingweb.eva.mpg.de/channumerals/

Greenberg, J. H. (1978). Generalizations about numeral systems. In J. H. Greenberg (Ed.), *Universals of human language* (Vol. 3, pp. 249–295). Stanford University Press.

[10] Klein (1992).

[11] Schlimm (2018).

Jeffreys, M. D. W. (1938). The cowry shell: A study of its history and use in Nigeria. *Nigeria, A Quarterly Magazine of General Interest, 15*(221–226), 256.

Jeffreys, M. D. W. (1948). The diffusion of cowries and Egyptian culture in Africa. *American Anthropologist, New Series, 50*(1), 45–53.

Kendall, T. (1817). A vocabulary of English and New Zealand words, the latter being compared with some of those in the Tonga language. In J. L. Nicholas (Ed.), *Narrative of a voyage to New Zealand, performed in the years 1814 and 1816, in company with the Rev. Samuel Marsden, principal chaplain of New South Wales* (Vol. II, pp. 327–352). James Black and Son.

Klein, J. (1992). *Greek mathematical thought and the origin of algebra*. Dover Publications.

Lee, S. (1820). *A grammar and vocabulary of the language of New Zealand*. Church Missionary Society.

Lemaitre, Y. (2004). Austral (Rurutu), French Polynesia. In E. Chan (Ed.), *Numeral systems of the world's languages*. Max Planck Institute. https://lingweb.eva.mpg.de/channumerals/Austral-Rurutu.htm

Nishimoto, N. (2015). Austral (Rurutu), French Polynesia. In E. Chan (Ed.), *Numeral systems of the world's languages*. Max Planck Institute. https://lingweb.eva.mpg.de/channumerals/Austral-Rurutu.htm

Overmann, K. A. (2020). The curious idea that Māori once counted by elevens, and the insights it still holds for cross-cultural numerical research. *Journal of the Polynesian Society, 129*(1), 59–84.

Schlimm, D. (2018). Numbers through numerals: The constitutive role of external representations. In S. Bangu (Ed.), *Naturalizing logico-mathematical knowledge: Approaches from psychology and cognitive science* (pp. 195–217). Routledge.

Chapter 11
Ngiti Numbers

Abstract The Ngiti are an ethnic group of the Democratic Republic of the Congo. Traditional Ngiti numbers were based on *thirty-two* with a sub-base of *four*. While this number system has largely fallen out of use and is at risk of disappearing, the Ngiti traditionally used it to count *bridewealth*, money and commodities that a groom's family pays to the bride's parents for the value of a woman to bear his children. Like systems of numbers based on counting cowries (as for example, the Yoruba and Igbo number systems), Ngiti numbers adjust bundles by *accumulating* (adding to a lower bundle) and *back-counting* (subtracting from a higher bundle). Accumulating, the most common strategy, is found in almost all number systems. Back-counting is much less frequent, as fewer number systems include it. Unusually, Ngiti numbers use a third strategy known as over-counting, which anticipates the completion of the set that intermediates between lower and higher bundles. Over-counting is relatively rare, found in only a handful of number systems: besides Ngiti, number systems that use over-counting include Kakoli (Papua New Guinea), Old Norse (Scandinavia), and Old Tagalog (the Philippines).

Keywords Duotrigesimal · Base-32 · Accumulating · Back-counting · Over-counting

The Ngiti (or Southern Lendu) of Zaire, today the Democratic Republic of the Congo (DRC), had a traditional number system based on *thirty-two* (base-32) with a sub-base of *four*.[1] It was already well on its way to being forgotten several decades ago, and today has largely been replaced by a system of decimal numbers:[2]

> Traditionally, Ngiti has a four-based numeral system which is totally in disuse nowadays. Only a few old people remember it. It survived longest in its use for important elements in social life: a) for the dowry system, b) in court cases, c) in wartime, for counting arrows in

[1] Hammarström (2010); Lojenga (1994).

[2] Lojenga (1994), pp. 355–356.

K. A. Overmann, *Cultural Number Systems*, Interdisciplinary Contributions to Archaeology, https://doi.org/10.1007/978-3-031-83383-0_11

> a quiver (32 arrows per quiver), and d) for the famous "bao" game, which has four rows of eight, i.e., 32 holes, and four grains for each of eight holes, i.e., 32 grains for each player. (Lojenga, 1994, pp. 355–356)

Let's take a closer look at the first social use, which is really *bridewealth* (or *bride-price*) rather than dowry; although the two terms are often used interchangeably, bridewealth is paid by the groom's family to the bride's parents for the value of a woman to bear his children, while a dowry is paid by the bride's family and accompanies her to the new household to bolster her status there.[3] In the DRC, marriage payments are made by the groom or his family to the parents of the bride.[4] In the past, these payments consisted of livestock (typically, goats or cattle) and household goods (blankets, clothing, utensils); today, bridewealth is more likely to include money, as well as commodities like "motorcycles, computers, smartphones and televisions."[5]

There aren't any records of how bridewealth was counted with traditional Ngiti numbers. However, it's possible to hypothesize based on counting methods used in other number systems. An example is counting with cowries in Yoruba. Small groups of cowries are formed and then combined as larger groups, with amounts intermediate to the completed bundles either incrementing from the closest lower bundle or decrementing from the closest higher bundle. The midpoint between the two bundles acts to minimize the number of items that must be moved. Thus, Yoruba numbers between 400 (*two* bundles of 200) and 600 (*three* bundles of 200) are formed by adding to 400 as the closest lower bundle or subtracting from 600 as the closest higher bundle, with 500 being the midpoint. Thus, 435 can be expressed as *irínwó o le marundilogoji*, *two* bundles of 200 plus *five* from *two* twenties, 525 as *òrìn dín légbèta ólé márùń*, *four* twenties from *three* bundles of 200 plus *five*.[6] A similar logic used to adjust Ngiti bundles.

Ngiti adjusts bundles by accumulating and back-counting. Cross-culturally, the most common strategies either add to the closest lower bundle or subtract from the closest higher one (Fig. 11.1). The former is accumulating, while the latter is known as *back-counting*.[7] We've seen back-counting previously as the Roman formation for *nineteen* (*undeviginti*, *one from twenty*), the Sumerian formation for 18 (*20 minus 2*),[8] and the Yoruba formation for *fifteen* (*ẹ̀ẹ́dogún*, *five from twenty*). In Roman numerals, the digit to be subtracted (the "subtrahend") was placed before the number it would be subtracted from (the "minuend"), so 9 (IX) was 1 from 10. In Sumerian numerals, a special sign meaning "minus" (la_2) separated minuend from subtrahend, so 47 (𒐐𒇲𒐈) was 3 set off from 50. Back-counting also occurs in finger-counting; an example is found in Sessuto numbers, where the names for *nine*

[3] Goody and Tambiah (1973).

[4] Bresnihan (2017).

[5] Goody and Tambiah (1973); Mutombo (2018).

[6] Armstrong (1962), p. 17.

[7] Menninger (1992).

[8] Friberg (2009); Friberg and Al-Rawi (2016); also see Menninger (1992).

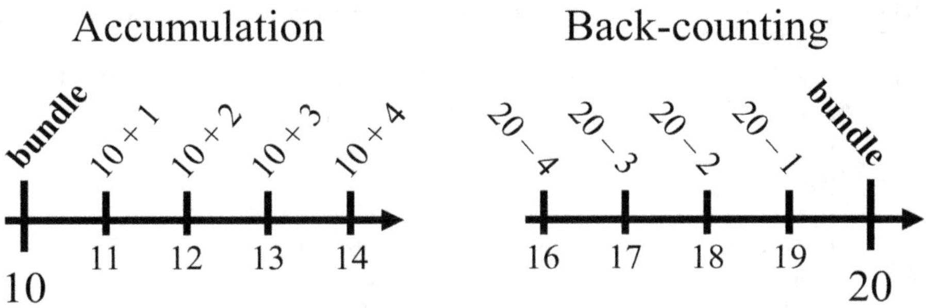

Fig. 11.1 Common strategies for adjusting bundles. (Left) Accumulation adds to the closest lower bundle, while (right) back-counting subtracts from the closest higher bundle. Image by the author

Table 11.1 Traditional Ngiti numbers are a base-32 system with accumulation and back-counting (Lojenga, 1994)

Nr.	Ngiti	Meaning	Nr.	Ngiti	Meaning
1	*atdí*	1	19	*àbà-vi*	20 – 1
2	*ɔyɔ*	2	20	*àbà*	20
3	*ɨbhʉ*	3	21	*àbà dʒ atdí*	20 + 1
4	*ɨfɔ*	4	22	*àbà dʒ ɔyɔ*	20 + 2
5	*imbo*	5	23	*àròtsí-vi*	24 – 1
6	*aza*	6	24	*àròtsí*	24
7	*àrʉ̀bhʉ̀*	8 – 1	25	*àròtsí dʒ atdí*	24 + 1
8	*àrʉ̀*	8	26	*àròtsí dʒ ɔyɔ*	24 + 2
9	*àrʉ̀gyètdí*	8 + 1	27	*àdzòro-vi*	28 – 1
10	*idrɛ*	10	28	*àdzòro*	28
11	*otsi-vi*	12 – 1	29	*àdzòro dʒ atdí*	28 + 1
12	*otsi*	12	30	*àdzòro dʒ ɔyɔ*	28 + 2
13	*otsi dʒ atdí*	12 + 1	31	*wǎdhɨ̀-vi*	32 – 1
14	*otsi dʒ ɔyɔ*	12 + 2	32	*wǎdhɨ̀*	32
15	*ɔpɨ-vi*	16 – 1	64	*ɔyɔ wǎdhɨ̀*	2 × 32
16	*ɔpɨ*	16	96	*ɨbhʉ wǎdhɨ̀*	3 × 32
17	*ɔpɨ dʒ atdí*	16 + 1	128	*ɨfɔ wǎdhɨ̀*	4 × 32
18	*ɔpɨ dʒ ɔyɔ*	16 + 2			

and *eight* describe both hands (*ten* fingers) minus the *one* or *two* fingers that are folded, rather than extended.[9]

Ngiti uses both of these strategies (Table 11.1). Starting with *twelve*, items are added to the closest lower bundle of *four* to produce the next two numbers, so *thirteen* is *twelve and one* and *fourteen* is *twelve and two*. The next number is back-counted, so *fifteen* is *sixteen lacking one*.

[9] Mabille (1893); Matsau (2009); Schrumpf (1862).

Table 11.2 Traditional Kakoli numbers are a base-24 system with over-counting (Bowers & Lepi, 1975; Head, 2008)

Nr.	Kakoli	Meaning	Nr.	Kakoli	Meaning
1	*telu*	1	19	*supuŋga yepoko*	3 to 20
2	*talu*	2	20	*supu*	20
3	*yepoko*	3	21	*tokapuŋga telu*	1 to 24
4	*kise*	4	22	*tokapuŋga talu*	2 to 24
5	*te(lu)pakara*	1 + 4	23	*tokapuŋga yepoko*	3 to 24
6	*talupakera*	2 + 4	24	*tokapu*	24
7	*yepokopakera*	3 + 4	25	*alapuŋga telu*	1 to 28
8	*eŋgaki*	8	26	*alapuŋga talu*	2 to 28
9	*rurepoŋga telu*	1 to 12	27	*alapuŋga yepoko*	3 to 28
10	*rurepoŋga talu*	2 to 12	28	*alapu*	28
11	*rurepoŋga yepoko*	3 to 12	29	*polaŋgipuŋga telu*	1 to 32
12	*rurepo*	12	30	*polaŋgipuŋga talu*	2 to 32
13	*malapuŋga telu*	1 to 16	31	*polaŋgipuŋga yepoko*	3 to 32
14	*malapuŋga talu*	2 to 16	32	*polaŋgipu*	32
15	*malapuŋga yepoko*	3 to 16	48	*tokapu talu*	24 × 2
16	*malapu*	16	72	*tokapu yepoko*	24 × 3
17	*supuŋga telu*	1 to 20	96	*tokapu kise*	24 × 4
18	*supuŋga talu*	2 to 20			

A third strategy for adjusting bundles anticipates their completion. Where back-counting subtracts from a completed larger bundle, over-counting anticipates its completion and adds toward it.[10] This strategy falls somewhere between forming bundles, as in the cowrie bundles of Yoruba, and using plural units, as in the tally-counting of Polynesia, where a bundle of multiple items is counted as *one*. When plural units are used, incomplete bundles are treated as a remainder (e.g., *one*, *two*, or *three* items when counting groups of *four*).[11] In over-counting, incomplete bundles are credited toward the next higher bundle. An interesting example comes from Old Norse, where *fjórir tigir manna annars hundraðs* meant *four tens of another hundred men*, or 160 men. *Hundraðs* is the *long hundred* valued as 120,[12] and the next *hundraðs* (or 240) is anticipated with *four tens* toward it, starting from 120. It's worth noting that no known system of written numerals represents over-counted numbers (we'll see this again in discussing Old Tagalog numbers in Chap. 24).

In the highlands of Papua New Guinea, the Kakoli (or Kaugel) language has a counting system that uses groups of *four* like Ngiti does (Table 11.2), but over-counts them (Fig. 11.2). Thus, *thirteen*, *fourteen*, and *fifteen* are *one*, *two*, and *three* toward *sixteen*. Where Ngiti forms its larger bundles on *eight* groups of *four*, creating a duotrigesimal (base-32) structure, Kakoli forms its larger bundles on *six*

[10] Menninger (1992).

[11] Overmann (2020).

[12] Gordon (1957), pp. 48 and 214.

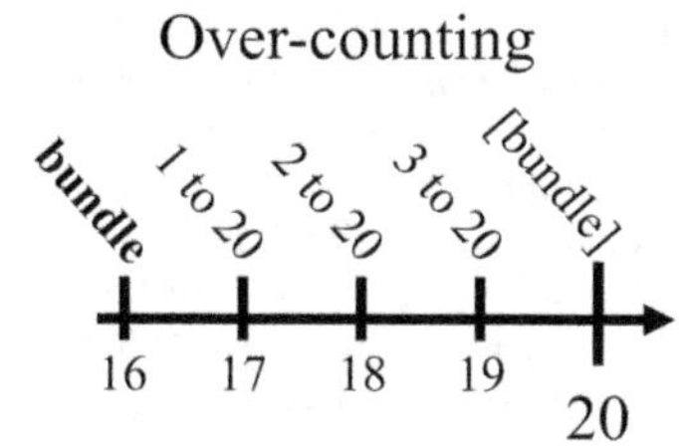

Fig. 11.2 Over-counting anticipates the closest higher bundle and adds toward it from the closest lower bundle. Image by the author

groups of *four*, creating a quadravigesimal (base-24) structure. Interestingly, Kakoli uses over-counting for its larger bundles as well: Bundles of *four* are counted to 32, at which point a bundle of 24 is declared, along with *eight* toward the next bundle of 24 (*tokapu poranimbe eŋgaki ŋgoli lekemo*).[13]

Why particular bundle sizes were used is a good question. Such parameters are often based on pragmatic considerations of counting physical objects. In Kakoli, valuable objects like "goldlip pearlshells, pigs, [and] game animals"[14] were formally and publicly counted as a preface to their distribution. The main donor would proceed along the gifts, which had been arranged in rows for display; he would tap the objects or the stakes to which live animals had been tied, counting by twos to form the groups of *four* shown in Table 11.2.[15] Groups of *four* may have arranged the gifts advantageously for public viewing; traditional bridewealth/dowry transfers can involve similar ceremonial displays.[16] This doesn't answer the question of why groups of *eight* (Ngiti) or *six* (Kakoli) were used to form larger bundles. Perhaps these numbers were related to the public distribution of goods. Like the use of *six* as the next higher bundle in Sumerian, the passage of time and the ongoing loss of knowledge mean we'll likely never know.

References

Armstrong, R. G. (1962). *Yoruba numerals*. Oxford University Press.

Bowers, N., & Lepi, P. (1975). Kaugel Valley systems of reckoning. *Journal of the Polynesian Society, 84*(3), 309–324.

Bresnihan, S. (2017). *Understanding women's lives in polygamous marriages: Exploring community perspectives in DRC*. Trōcaire/Irish Aid.

Friberg, J. (2009). A geometric algorithm with solutions to quadratic equations in a Sumerian juridical document from Ur III Umma. *Cuneiform Digital Library Journal, 3*, 1–27.

Friberg, J., & Al-Rawi, F. N. H. (2016). *New mathematical cuneiform texts*. Springer.

Goody, J., & Tambiah, S. J. (1973). *Bridewealth and dowry*. Cambridge University Press.

Gordon, E. V. (1957). *An introduction to Old Norse* (2nd rev.). Clarendon Press.

[13] Bowers and Lepi (1975).

[14] Bowers and Lepi (1975), p. 312.

[15] Bowers and Lepi (1975), pp. 312–313.

[16] Goody and Tambiah (1973).

Hammarström, H. (2010). Rarities in numeral systems. In J. Wohlgemuth & M. Cysouw (Eds.), *Rethinking universals: How rarities affect linguistic theory* (pp. 11–59). Walter de Gruyter.

Head, J. (2008). Umbu-Ungu, Papua New Guinea. In E. Chan (Ed.), *Numeral systems of the world's languages*. Max Planck Institute. https://lingweb.eva.mpg.de/channumerals/Umbu-Ungu.htm

Lojenga, C. K. (1994). *Ngiti: A central-Sudanic language of Zaire*. Rüdiger Köppe.

Mabille, A. (1893). *Se-Suto-English and English-Se-Suto vocabulary*. Khatiso.

Matsau, M. A. (2009). Southern Sotho, Lesotho. In E. Chan (Ed.), *Numeral systems of the world's languages*. Max Planck Institute. https://lingweb.eva.mpg.de/channumerals/Sotho-Southern.htm

Menninger, K. (1992). *Number words and number symbols: A cultural history of numbers* (P. Broneer, Trans.). Dover Publications.

Mutombo, F. M. (2018). Gone are the goats: Wedding dowries get a tech upgrade in DRC. *Global Press Journal*. https://globalpressjournal.com/africa/democratic-republic-of-congo/gone-goats-wedding-dowries-get-tech-upgrade-drc/

Overmann, K. A. (2020). The curious idea that Māori once counted by elevens, and the insights it still holds for cross-cultural numerical research. *Journal of the Polynesian Society, 129*(1), 59–84.

Schrumpf, C. (1862). Sessuto. Ein Beitrag zur Süd-Afrikanischen Sprachenkunde, geschöpft aus meiner sechszehnjährigen Erfahrung [Sessuto. A contribution to South African linguistics, drawn from my sixteen years of experience]. *Zeitschrift der Deutschen Morgenländischen Gesellschaft, 16*(3), 448–481.

Chapter 12
Numbers in Sessuto

Abstract Sessuto is the language of the Basuto, an ethnic group of Lesotho, a sovereign kingdom landlocked within the larger country of South Africa. As recorded in the nineteenth century, the Basuto developed an innovative method of counting collaboratively. In this method, one person counted the units, a second person kept track of the tens, and a third person kept track of the hundreds. The method was extensible to very high numbers, as a fourth person would keep track of the thousands. This method of counting has also been found in Oceania. Early European observers assumed that collaborative counting meant the Basuto had difficulty with large numbers. In actuality, the method created an *ephemeral abacus*, ephemeral because the abacus existed only as long as the counting was performed and an abacus because the method represented the exponential values (units, tens, hundreds) in different places. The method simplified the task of counting for participants, who only needed to count from *one* to *ten* as assisted by their fingers. The method shows how higher numbers and attributes like place value and exponential structure can develop in the absence of formal devices and notations.

Keywords Ephemeral abacus · Collaborative counting · Place value · Exponential structure · Conventionalization

Sessuto (or Sotho) is the language of the Basuto people, a Bantu ethnic group in Lesotho, a sovereign kingdom landlocked within the larger country of South Africa. Sessuto not only had a term for *million*, the Basuto also developed an innovative way of counting collaboratively that shows how high numbers and attributes like place value and exponential structure can develop in the absence of formal devices and notations.

Counting beyond one hundred required the combined efforts of three men. This was documented by Christian Schrumpf (1818–1884), the Paris evangelical who arrived in Lesotho in 1842, founded a mission in Bethesda in 1843, and served there

---

A shorter version of this chapter originally appeared in Overmann (2023), and the material is reused here with permission.

K. A. Overmann, *Cultural Number Systems*, Interdisciplinary Contributions to Archaeology, https://doi.org/10.1007/978-3-031-83383-0_12

until his 1858 return to France because of poor health.[1] As the Rev. Schrumpf noted in a book he wrote about his 16 years of living and working among the Basuto,

> The adjectival numerals in the Sessuto language are extensive and somewhat awkward. That is why counting, when the number of objects to be counted is considerable, is almost a huge thing for the natives.
>
> When counting beyond a hundred, three men usually have to do the difficult work together. One man counts the units on his fingers, which he raises one after the other, thereby indicating the object counted or, if possible, touching it. The second man raises his fingers for the tens (always starting with the little finger of the left hand and continuing to the little finger of the right hand) as they are completed. The third man counts the hundreds.
> (Schrumpf, 1862, pp. 463–464, as translated)

It wasn't difficulty with large numbers. Schrumpf, along with other authors of the time, thought that this counting behavior indicated a particular awkwardness with numbers, perhaps the "difficulty of visualizing larger numbers by using the fingers only."[2] This interpretation is already familiar to us as the historical reaction to the method that Xerxes and Darius reportedly used to count their armies.[3]

In this case, it's not unreasonable to think that the average Basuto individual might have found it difficult to count from *one* to *one hundred* sequentially and by himself, simply because this wouldn't have been the usual practice. There are two other reasons that counting might have been difficult. First, the Sessuto names for numbers higher than *ten* were lengthy and fairly cumbersome descriptions of the different fingers involved. Counting from just *fifteen* to *twenty* was a lot to say: *leshume le metso e metlanu* ("a ten which has *five* roots"), *leshume le metso e tseletseng* ("a ten which has *six* roots"), *leshume le metso e shupileng* ("a ten which has *seven* roots"), *leshume le metso e robileng menomeli* ("a *ten* and a[nother] *ten* which breaks *two* fingers"), *leshume le metso e robileng mono o le mong* ("a *ten* and a[nother] *ten* which breaks *one* finger"), and *mashume a le mubeli* ("two tens").[4] The same numbers in German, the language Schrumpf used for his book, are much shorter—*fünfzehn*, *sechzehn*, *siebzehn*, *achtzehn*, *neunzehn*, and *zwanzig*—despite German's well-deserved reputation for lengthy words. Simply, the longer the names, the less suited they are for being recited to count things. Not only do longer names take more time to say, they also increase the burden that verbal counting puts on our finite capacities of memory and attention. That is, the longer the name, the greater the demands on our limited cognitive resources because there is more to remember and longer intervals of time between items.

Second, once counting has expended all the fingers (or segments or spaces, if used), counting to higher numbers requires some way of keeping track of the cycles and their multiples (in this case, the tens, hundreds, and so on). This too puts a burden on our memory and attention, which is why we so often use our fingers in the first place. Cross-culturally, it's why counting involves the hands of other people

---

[1] Jousse (1889), p. 345; Rosenthal (1978), p. 48.

[2] Cajori (1899), p. 35.

[3] Flower (2007); Grote (1851); Macan (1908a, 1908b); Maurice (1930); Rolfe (1946).

[4] Schrumpf (1862), pp. 464–465, as glossed in Matsau (2009); also see Mabille (1893), pp. 64–65.

once the original set of fingers has been used. In some number systems, the first person counts from *one* to *ten*, the second counts from *eleven* to *twenty*, the third from *twenty-one* to *thirty*, and so on. In Sessuto and some Oceanic number systems[5] (and likely many more that weren't as well documented), the first person counts to *ten*, the second counts the tens as the first person reaches them, the third counts the hundreds as the second person reaches them, and so on. The strategy is extensible, potentially to very high numbers, by adding more people to keep track of the thousands, ten thousands, and so on. Sessuto has a term for *million* (*sekete sa likete*),[6] so it's possible that three men counting to a *hundred* was just the beginning. Regardless of how many people become involved (and it would take six to count to a *million*), the task remains simple: Each person just counts from *one* to *ten* and keeps track with his or her fingers.

Thinking that the method of counting was just awkwardness misses the point. By keeping track of the units, tens, and hundreds, the Basuto created an abacus. As abacuses go, it was ephemeral, consisting not of a solidly physical device but of fingers and roles, and then, existing only so long as counting was being performed. It also was not well suited for calculating, since it would be difficult to move amounts around in the ways that calculation requires. The result was nonetheless a positional system, one where each man's literal position determined the value of the numbers he tracked.[7]

Collaborative counting had other benefits. For one thing, it made the exponential structure very clear. A number like 888 was expressed as *eight hundreds* (as counted by the third man), *eight tens* (as counted by the second man), and *eight units* (as counted by the first man). For another, sequential recitation, even just the words from *one* to *ten*, put longer phrases (like *eight* and *nine*) under pressure to shorten. Over time, cumbersome phrases like *e robileng meno e le meli* (*eight*, "a ten which breaks two fingers" or all the fingers of both hands with two fingers folded down[8]) would become unanalyzable syllables, words that no longer revealed their origin in finger-counting. In addition, collaborative counting would expose participants and witnesses alike to the naming of very high numbers by just the words for *one* through *ten* and productive terms for the tens, hundreds, and thousands. Such exposure would help the names to become conventionalized, and the process of naming to become lexicalized as well.

Historical descriptions like the one Schrumpf provided shouldn't be discounted. Cultural observers like Schrumpf, as products of the nineteenth century or earlier, were generally untrained and often biased, misunderstanding what they saw and characterizing it as primitive. Schrumpf certainly didn't appreciate the collaborative finger-counting he witnessed for the ephemeral abacus it created, despite his long experience with Basuto culture. Modern numerical researchers often discount or

---

[5] Churchward (1941); Collocott (1925, 1927).

[6] Schrumpf (1862).

[7] Gerdes (2008), p. 1762.

[8] Mabille (1893), pp. 64–65.

ignore such historical records. This is a mistake because, however imperfect, these records nonetheless provide valuable insight into what cultural number systems were like before they were much altered by their exposure to Western numbers. Historical descriptions are also often comparable cross-culturally, like the collaborative strategy used in Sessuto can be compared to the ones described for Oceania, and this provides a measure of validity. Comparability isn't a matter of the cultures involved being necessarily connected or related in some way; rather, different societies solve similar problems with the same set of resources, producing similar and comparable solutions. We see this a lot in number systems, which are strongly similar despite being widely separated by distance, language, culture, and time.

References

Cajori, F. (1899). The number concept. *Colorado College Studies: Papers Read before the Colorado College Scientific Society, 8*, 32–40.

Churchward, C. M. (1941). *A new Fijian grammar*. Australasian Medical Publishing Company.

Collocott, E. E. V. (1925). Supplementary Tongan vocabulary; also notes on measuring and counting, proverbial expressions and phases of the moon. *Journal of the Polynesian Society, 34*(3), 193–213.

Collocott, E. E. V. (1927). Kava ceremonial in Tonga. *Journal of the Polynesian Society, 36*(1), 21–47.

Flower, M. A. (2007). The size of Xerxes' expeditionary force. In R. B. Strassler (Ed.), *The landmark Herodotus: The histories* (pp. 819–823). Pantheon Books.

Gerdes, P. (2008). Number theory in Africa. In H. Selin (Ed.), *Encyclopaedia of the history of science, technology, and medicine in non-Western cultures* (2nd ed., pp. 1761–1763). Springer.

Grote, G. (1851). *History of Greece* (Vol. V). John P. Jewett and Company.

Jousse, T. (1889). *La mission française évangélique au sud de l'Afrique son origine et son développement jusqu'a nos jours [The French evangelical mission in South Africa: Its origin and development to the present day]* (Vol. 2). Librairie Fischbacher.

Mabille, A. (1893). *Se-Suto-English and English-Se-Suto vocabulary*. Khatiso.

Macan, R. W. (1908a). *Herodotus: The seventh, eighth & ninth books. Vol. I, Part 1: Introduction, Book VII (text and commentaries)*. Macmillan.

Macan, R. W. (1908b). *Herodotus: The seventh, eighth & ninth books. Vol. II: Appendices, indices, maps*. Macmillan.

Matsau, M. A. (2009). Southern Sotho, Lesotho. In E. Chan (Ed.), *Numeral systems of the world's languages*. Max Planck Institute. https://lingweb.eva.mpg.de/channumerals/Sotho-Southern.htm

Maurice, F. (1930). The size of the army of Xerxes in the invasion of Greece 480 B.C. *Journal of Hellenic Studies, 50*(2), 210–235.

Overmann, K. A. (2023). *The materiality of numbers: Emergence and elaboration from prehistory to present*. Cambridge University Press.

Rolfe, J. C. (1946). *Quintus Curtius* (Vol. I). Harvard University Press.

Rosenthal, E. (1978). *Encyclopaedia of Southern Africa* (7th ed.). Juta and Company Ltd.

Schrumpf, C. (1862). Sessuto. Ein Beitrag zur Süd-Afrikanischen Sprachenkunde, geschöpft aus meiner sechszehnjährigen Erfahrung [Sessuto. A contribution to South African linguistics, drawn from my sixteen years of experience]. *Zeitschrift der Deutschen Morgenländischen Gesellschaft, 16*(3), 448–481.

Chapter 13
African Finger-Counting

Abstract Several interesting finger-counting systems were documented in Africa in the nineteenth and early twentieth centuries. In some, gestures were used instead of number-words. In others, gestures were accompanied by words, either the names of the numbers displayed by the fingers or expressions drawing attention to the fingers as they displayed numbers. Occasionally the number-word and accompanying gesture did not match, with the gesture providing the intended meaning. These uses suggest that finger-counting precedes names for numbers and helps lengthy names for numbers to shorten to unanalyzable words. Some finger-counting is symmetrical, involving the same number of fingers on both hands to express numbers (for example, *six* would be displayed as three fingers on each hand), while other finger-counting is sequential, using all the fingers on one hand before adding fingers on the other (in this case, *six* would be displayed as five fingers on one hand and one finger on the other). The Maasai use conventional gestures, finger-patterns for numbers whose meaning cannot be understood as the number of fingers displayed; conventions enable this system to express very high numbers.

Keywords Analyzability · Etymological transparency · Symmetrical · Sequential · Gesture

Besides the collaborative finger-counting of Sessuto and its inherent place value, some very interesting African finger-counting systems were documented between the nineteenth and early twentieth centuries by various missionaries, colonial administrators, and travelers. Here are just a few.

In some systems, gesture was used instead of words. Examples were noted by the German physician and world-traveler Henry Lichtenstein in the early nineteenth century.[1] Of the Xhosa, a people of South Africa, he wrote that sometimes "the numeral is not mentioned; the action of raising the fingers only is used"; of the Tswana of Botswana, he said, "The numbers are commonly expressed … by fingers

[1] Lichtenstein (1812).

K. A. Overmann, *Cultural Number Systems*, Interdisciplinary Contributions to Archaeology, https://doi.org/10.1007/978-3-031-83383-0_13

held up, so that the word is rarely spoken: many are even unacquainted with these numerals, and never employ any thing but the sign."[2] Among the Baluba, a people of the Democratic Republic of the Congo (DRC), gestures always accompanied the verbal expression of a number.[3] Gesture enables people to express numbers when there isn't a common language between the communicating parties, or when the communicants either don't have any words for numbers in their languages or don't know the ones that might be available, or when the established vocabularies for numbers consist of lengthy, cumbersome phrases that translate along the lines of *all the fingers on one of my hands and two of the fingers from my other hand.*

Any words that are used in conjunction with expressing numbers might have served merely to draw attention to the hands. An example comes from the Nkosi language, a Bantu dialect spoken in Cameroon. As recorded in the early twentieth century by the German missionary Heinrich Dorsch, speakers didn't verbally name the numbers they were communicating; instead, they introduced them by saying "nä" while displaying the appropriate number of fingers.[4] In other cases, the words used might be an indeterminate phrase meaning something like "this one" or "and another." In eastern Namibia, for example, the expression *neba hawu* ("this is finished") was used alike for the numbers *seven*, *eight*, *nine*, and *ten*; the verbal phrase depended on the requisite fingers being displayed to express the number meant.[5] In a few cases, the verbal expression was numerically misleading, and meaning depended on the fingers being used. An example was the phrase *tʃinɛ lubali*, which meant *four on one side* and was used to express the number *seven* by the Bashila people of the Democratic Republic of the Congo (DRC). Although the verbal expression by itself didn't readily express the idea of *seven*, it was accompanied by a gesture that displayed four fingers on one hand and three on the other.[6]

Using the fingers can support the shortening of lengthy phrases. In the part of the DRC once known as Zaire, a term for *nine* that originally meant *there is missing one out of ten*, *pabula kimɔ kia likumi*, could be shortened to simply *pabula*, *there is missing*.[7] The numerical meaning of the verbal expression was supported, despite its truncation, by the accompanying display of the fingers. This suggests that short, unanalyzable words like *nine* can emerge from lengthy phrases that describe the various arrangements of different fingers because their meaning is supported and perpetuated by the visual display that the fingers provide.

Finger-counting can be symmetrical, rather than sequential. In many finger-counting systems, including ones commonly used in the Western tradition, the fingers are used sequentially. For example, counting might proceed from thumb to little finger (or vice versa) and then transition to the other hand, where the same

[2] Lichtenstein (1812), Appendix.

[3] Gabriel (1921), p. 17.

[4] Dorsch (1910), pp. 256–257.

[5] Olderogge (1982).

[6] Tempels (1938).

[7] Tempels (1938), p. 52.

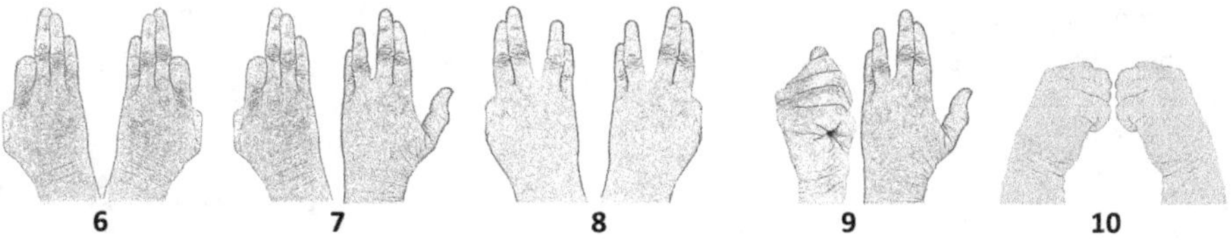

Fig. 13.1 Symmetrical finger-counting (*six*, *eight*, and *ten*) in the Bashila system. Image by the author after Tempels (1938, p. 51)

sequence is repeated or reversed.[8] In such systems, numbers like *six* and *eight* typically consist of *five* fingers on one hand and *one* or *three* fingers on the other. In some of the African systems, however, these numbers are instead presented symmetrically, as *three* fingers on both hands (*six*) or *four* fingers on both hands (*eight*). This is true of the finger-counting documented for the Bashila, a people from the Lake Mweru region between the DRC and Zambia (Fig. 13.1). The logic is twofold: a doubling of the base number, with *six* being *twice three* and *eight* being *twice four*, and a symmetrical arrangement of the fingers being used.

Authors like the Soviet anthropologist Dmitri Olderogge have suggested that such finger arrangements may have once represented binary counting, or counting things by means of pairing them.[9] For example, in the system of finger-counting used by the Bemba people of Zambia, the number *four* (Fig. 13.2) was displayed as four fingers on one hand, separated down the middle into two groups of *two*. Western readers might envision this as being akin to the hand sign that accompanied the Vulcan greeting "live long and prosper" in the *Star Trek* series.

In some systems, numerical gestures are conventional. Rather than instantiating number through the quantity of fingers, like five fingers would express *five*, the quantity of fingers in a conventional gesture are not related to the number being expressed. For example, the Maasai, a people of Kenya and Tanzania, count from *one* to *nine* as shown in Fig. 13.3 below.[10] While the gestures for the numbers *one* and *two* express quantity by extending and displaying one or two fingers, the gestures for the numbers *three* and higher do not. Interestingly, these same finger-signs remain in use today.[11]

Systems with conventional gestures for numbers tend to be associated with counting to relatively high numbers, often in the hundreds or thousands. For example, the Maasai expressed the hundreds by opening and closing the hand once (100) or twice (200).[12] The Arusha, a people of Tanzania, expressed the thousands by repeating the sign for *hundred*, which involved the fist of the right hand being

[8] Domahs et al. (2012); Pika et al. (2009); also see Overmann (2014, 2021).

[9] Olderogge (1982, 1984).

[10] Merker (1904).

[11] Yadav (2021).

[12] Thomson (1886).

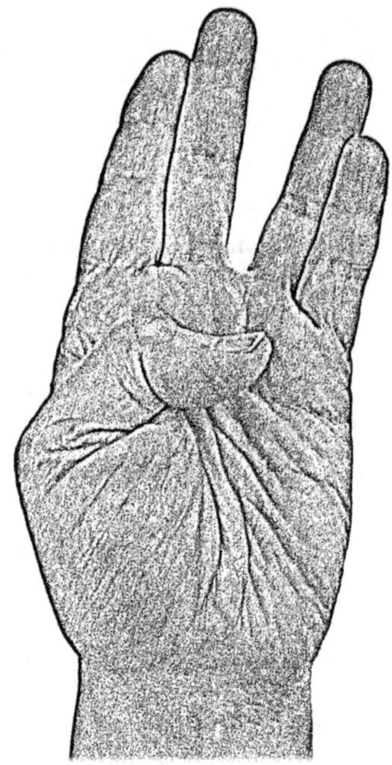

Fig. 13.2 The Bemba sign for the number *four*. Image by the author after Olderogge (1982, Fig. 4, p. 20)

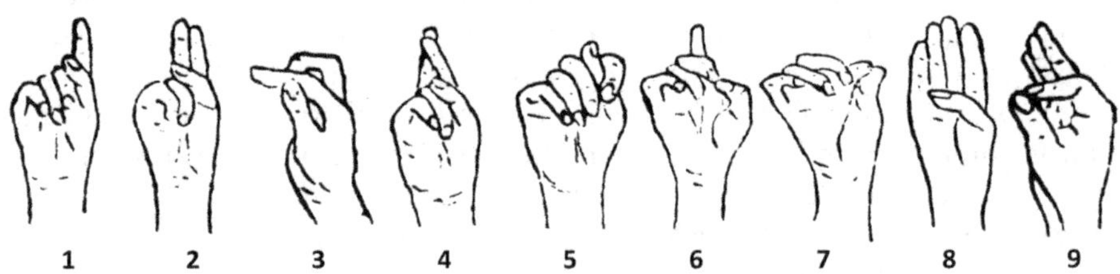

Fig. 13.3 Maasai finger-counting from *one* to *nine*. Adapted from Merker (1904, Fig. 73, p. 150). Image in the public domain

"loosely clenched and then fingers shot out and splayed apart," *five* times for *thousand* and *ten* times for *two thousand*.[13] While the repetition was numerically meaningful, the multiplication wasn't exact, since *five* times a *hundred* is *five hundred*, not a *thousand*; this too was apparently a matter of convention, since there was a separate sign for *five hundred*: the sign for *five*, a clenched fist with the tip of the thumb protruding between the first and second fingers, made while twisting the wrist.[14] The arbitrary connections between the gestures and the numbers they

[13] Gulliver (1958), pp. 259–260.

[14] Gulliver (1958), pp. 259–260.

express suggest that the number systems using such gestures may be quite old, evoking the dactylonomy (digital reckoning) of Europe and the Near East (Chap. 19). The gestures have become like written symbols for numbers that are learned conventions rather than signs that convey number through the quantity of their elements (e.g., 7 vs. 𒐛).[15]

References

Domahs, F., Kaufmann, L., & Fischer, M. H. (Eds.). (2012). *Handy numbers: Finger counting and numerical cognition*. Frontiers Media SA.

Dorsch, H. (1910). Grammatik der Nkosi-Sprache mit einer der Nkosi mit Duala vergleichenden Einleitung [Grammar of the Nkosi language with an introduction comparing Nkosi with Duala]. *Zeitschrift für Kolonialsprachen, 1*, 241–283.

Gabriel, F. (1921). *Étude du Tshiluba [Study of Tshiluba]*. Ministère des Colonies.

Gulliver, P. H. (1958). Counting on the fingers by two East African tribes. *Tanganyika Notes and Records, 51*, 259–262.

Lichtenstein, H. (1812). *Travels in Southern Africa, in the years 1803, 1804, 1805, and 1806* (A. Plumptre, Trans., Vol. 1). Henry Colburn.

Merker, M. (1904). *Die Masai: Ethnographische Monographie eines ostafrikanischen Semitenvolkes [The Masai: Ethnographic monography of an East African Semitic people]*. Dietrich Reimer.

Olderogge, D. A. (1982). СИСТЕМЫ СЧЕТА В ЯЗЫКАХ НАРОДОВ ТРОПИЧЕСКОЙ И ЮЖНОЙ АФРИКИ [Counting systems in the languages of the peoples of tropical and South Africa]. *Trudy Instituta Etnografii, 13*, 3–33.

Olderogge, D. A. (1984). СИСТЕМЫ СЧЕТА В ЯЗЫКАХ НАРОДОВ ТРОПИЧЕСКОЙ И ЮЖНОЙ АФРИКИ [Counting systems in the languages of the peoples of tropical and South Africa]. *Trudy Instituta Etnografii, 14*, 3–16.

Overmann, K. A. (2014). Finger-counting in the Upper Palaeolithic. *Rock Art Research, 31*(1), 63–80.

Overmann, K. A. (2021). Finger-counting and numerical structure. *Frontiers in Psychology, 12*, 723492.

Overmann, K. A. (2022). Early writing: A cognitive archaeological perspective on literacy and numeracy. *Visible Language, 56*(1), 8–44.

Pika, S., Nicoladis, E., & Marentette, P. (2009). How to order a beer: Cultural differences in the use of conventional gestures for numbers. *Journal of Cross-Cultural Psychology, 40*(1), 70–80.

Tempels, P. (1938). De telgebaren der Bashila [The messages of Bashila]. *Congo-Overzee, IV*(2), 49–53.

Thomson, J. (1886). Sign-numbers in use among the Masai. *Nature, 34*, 522.

Yadav, V. (2021). Count like a Maasai! Counting with hands. *Meet the Maasai*. Project FUEL. https://artsandculture.google.com/story/count-like-a-maasai-project-fuel/xgVRfDalLnOdAg?hl=en

[15] Overmann (2022).

Part III
Number Systems in Europe

Chapter 14
Prehistoric Finger-Counting

Abstract Cross-culturally, fingers are the first material structure used for counting because they are neurologically integrated with the ability to perceive quantity, visually salient, socially prevalent, and standard in their quantity. Since gestures don't preserve, finger-counting tends to be archaeologically invisible. A possible exception comes from France. Around 25,000 years ago at Cosquer and Gargas, people placed their hands against the cave walls and then blew paint around them, leaving outlines of their fingers and palm. Some of these stencils have shortened fingers. While scholars have speculated on possible causes, the sets of finger-patterns at these caves are not consistent with disease, injury, or ritual mutilation. Instead, the Cosquer set looks like finger-counting performed by sequentially extending the fingers of one hand. The Gargas set contains the same five patterns, along with five more that might have been hunting signs. Nonetheless, there are reasons to be cautious about this interpretation: for example, paint degradation makes finger length difficult to judge. Traits like color, placement within the cave, and the makers' age and gender also mean that the stencils' social purpose remains a mystery.

Keywords Cosquer · Gargas · Finger-counting · Cave painting · Negative stencil

All human societies that count do so by involving their fingers, regardless of how old or complex their number system might happen to be. We can see this in Western societies, where finger-counting coexists with a number system that is tens of thousands of years old and highly complex. Fingers also appear to be the very first material structure that all societies use for counting, not only because the fingers are neurologically interconnected with our ability to perceive quantity, but also because they are ubiquitous and standard (most people have ten of them), handy (quite literally at our fingertips), and individuable (as the little finger differs from the ring finger adjacent to it). So, we might ask how long humans have been counting with their fingers. This question is challenging because finger-counting isn't the type of thing that tends to leave traces of itself behind. However, some of Europe's prehistoric caves may hint at an answer.

K. A. Overmann, *Cultural Number Systems*, Interdisciplinary Contributions to Archaeology, https://doi.org/10.1007/978-3-031-83383-0_14

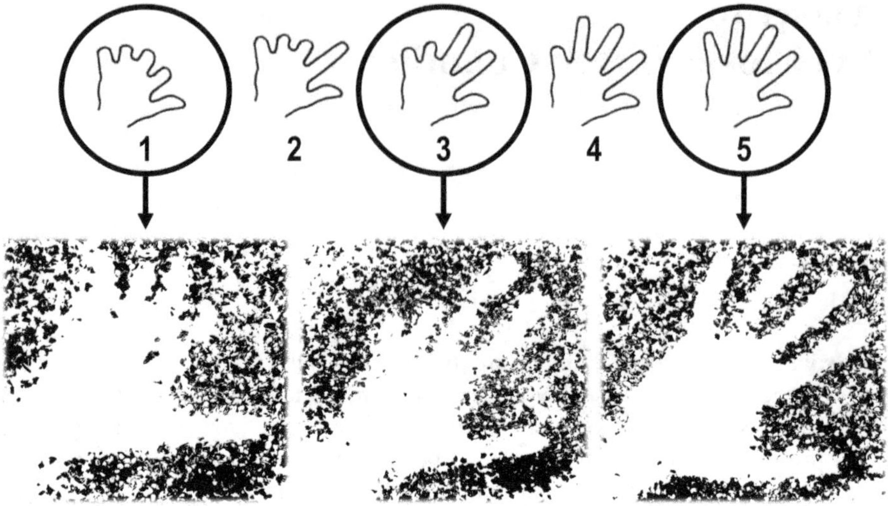

Fig. 14.1 Finger-counting patterns (modern reconstructions) resembling those at Cosquer Cave

People may have counted with their fingers in the Upper Palaeolithic. Cosquer is a cave on the southern coast of France. Some 25,000 years ago, people decorated the walls of the cave with paintings of animals and negative stencils of their hands. A negative stencil is made by placing the hand on the cave wall and then blowing or dabbing paint around it; a positive stencil, in comparison, is formed by dipping the hand in paint and then touching it to the wall. Cosquer contains more than sixty negative hand stencils, making it one of the largest collections of prehistoric hand stencils in all of Europe.[1] About two-thirds of the Cosquer stencils were made with a black substance, while the others were made in red.[2]

As is common in Palaeolithic stencils, many of the Cosquer stencils have shorter-than-normal fingers. For the past century or so, these shortened digits were thought to have represented fingers lost through frostbite, disease, accidents, or mutilation rituals.[3] While debate over the reason for the shortened digits remains ongoing,[4] a single hand is able to make 32 different patterns with combinations of long and short fingers, from all five fingers extended to all five bent. In 2006, an analysis of 49 of the Cosquer stencils found that they comprised only five of the 32 possible patterns (Fig. 14.1).[5] So, the immediate question we must ask is this: why only these five and not the other 27?

[1] Clottes et al. (1992); Pike et al. (2012).

[2] Clottes and Courtin (1994); Clottes et al. (2005).

[3] Leroi-Gourhan (1967); Luquet (1926, 1938).

[4] E.g., Alonso (2019); McCauley et al. (2018).

[5] Rouillon (2006).

The Cosquer stencils look like finger-counting. All known finger-counting systems express numbers through different combinations of fingers (though some additionally involve the toes and other body parts). A common (but hardly universal) sequence involves extending fingers from one side of the hand to the other in sequential order.[6] For example, *one* might be represented by extending just the thumb, *two* by extending the thumb and index finger, *three* by the thumb and index and middle fingers, *four* by all but the little finger, and *five* by all five fingers. Interestingly, these are the exact five patterns found at Cosquer.[7]

Some finger patterns are more difficult to make than others because of the way our hand muscles are configured.[8] For example, many people find that bending the little finger while extending the ring finger is difficult because these fingers share a common muscle, which makes independent movement challenging. This so-called infeasible pattern is one of the five patterns found at Cosquer, the one we are calling *four*. Its inclusion suggests that the people who made the stencils prioritized the sequence it helped form over the ease of its production.[9]

Nearby in the French Pyrenees is Gargas Cave, a site that is a little bit older than Cosquer at 27,000 years old. Like Cosquer, the 112 stencils in this cave were analyzed in terms of their digit lengths. Of the 32 possible patterns, the Gargas stencils contain ten: the same five that look like finger-counting at Cosquer (87 stencils) and another five that may have represented hunting signals (25 stencils).[10] The fact that the same subset of the 32 possible patterns occurs in more than one painted cave suggests that the finger patterns were indeed meaningful to the societies who decorated the caves.[11] However, this still doesn't mean that they were necessarily meaningful as numbers.

If we interpret them numerically, the stencils at Cosquer and Gargas suggest a number system capable of counting no higher than *five*. Both caves contain stencils with *one*, *two*, *three*, *four*, and *five* long fingers. The stencil for *four* occurs less frequently, and no combination of stencils suggests any numbers higher than *five*. These characteristics are consistent with the way that numbers emerge in contemporary number systems: The quantities that we are able to perceive and appreciate—*one*, *two*, and perhaps *three* and occasionally *four*—emerge as numbers first, followed by the number *five* as the number of fingers on one hand.[12] The number *four* is most typically a "fuzzy number" that becomes discrete or a "gap number" that gets filled in because there is a finger on the hand to represent it, the one falling between the fingers representing *three* and *five*.[13]

[6] Overmann (2014, 2021).

[7] Rouillon (2006).

[8] Lin et al. (2000).

[9] Overmann (2014).

[10] Leroi-Gourhan (1967).

[11] Etxepare and Irurtzun (2021).

[12] Overmann (2014).

[13] Overmann (2023).

The stencils were unlikely to have been made by hands with missing fingers. Accidents, disease, and ritual mutilation would produce combinations of short and long digits different from those seen at Cosquer and Gargas. Finger loss from accidents and disease reflects the use of the hands: Because the thumb and index and middle fingers are primarily involved in tool use, they are the most likely to be injured. Yet the thumb is present in all of the Cosquer and Gargas stencils, and the majority of stencils additionally contain the index and middle fingers. Conversely, ritual mutilation tends to shorten the little and ring fingers because their removal has the least effect on the hand's remaining strength and usability. These patterns (*three* and *four* in Fig. 14.1) are among the five that look like finger-counting; the little and ring fingers appear in about half the stencils at Cosquer and about a fifth of those at Gargas. Significantly, removal of the index and middle fingers (patterns *one* and *two*) creates a severely mutilated hand. A hand with only the thumb remaining would surely present a significant challenge to surviving the rugged prehistoric lifestyle.

Other stencil characteristics suggest non-numerical meanings. The hand stencils are only a small portion of the cave art at Cosquer and Gargas, and the complexity of the total decoration suggests the that societies who made them were complex and likely wealthy.[14] Certainly, the societies could support artists who painted the images and specialists who led the associated rituals. Complexity and wealth reliably precede and motivate numerical elaboration—things like counting to higher numbers. This is because numbers are a good technology for keeping track of things, and they develop in response to having things to be kept track of.[15] Nonetheless, wealth and complexity do not guarantee that numerical elaboration has actually occurred, just that the conditions are right for it to develop.

There are reasons to be cautious about a numerical interpretation. Some of the Cosquer stencils were omitted from the 2006 study because they were too faint for their digit lengths to be discerned.[16] The substances used as prehistoric paints degrade over time because of moisture and other natural conditions found in caves. If the original digit lengths could be identified, we might find that they didn't fit the same pattern. The size of the hands also suggests that many of the stencils were made by women or children,[17] and the social meaning and importance of this are difficult for us to judge. The stencils not only differ in their color (black or red), they also vary in terms of which hand was used (left or right), how they were oriented (upright or sideways), how frequently they were repeated, and whether they appeared by themselves, with other hand stencils, or with painted animal images. They also differ in where they were placed inside the cave; some cave locations are

[14] Hayden (2021); Owens and Hayden (1997).

[15] Overmann (2013).

[16] Rouillon (2006).

[17] Clottes et al. (2008); Snow (2013).

more accessible and/or visible than others, and this likely had a cultural meaning as well.[18]

Even if the stencils had a numerical intent, their other characteristics have the potential for social meanings and purposes with nothing whatsoever to do with numbers. If the stencils were meant to be numbers, they are certainly not unambiguous as such to our eyes. That is, if one of the patterns meant *three*, we don't understand the purpose or meaning of repeatedly stenciling *three* in red on a particular wall of the cave. So, the stencils continue to keep their secrets about their meanings and purposes within prehistoric life.

References

Alonso, J. R. (2019). Missing fingers. *European Journal of Anatomy, 23*(2), 147–149.

Clottes, J., & Courtin, J. (1994). *La grotte Cosquer: Peintures et gravures de la caverne engloutie [Cosquer Cave: Paintings and engravings of the submerged cavern]*. Seuil.

Clottes, J., Courtin, J., Chazine, J.-M., & Noury, A. (2008). *Hands stencils in Cosquer Cave: Men and/or women?* Paper presented to the sixth World Archaeological Congress, held in Dublin, Ireland from 29th June to 4th July 2008.

Clottes, J., Courtin, J., & Vanrell, L. (2005). *Cosquer redécouvert [Cosquer rediscovered]*. Seuil.

Clottes, J., Valladas, H., Cachier, H., & Arnold, M. (1992). Des dates pour Niaux et Gargas [Dates for Niaux and Gargas]. *Bulletin de la Société Préhistorique Française, 89*(9), 270–274.

Etxepare, R., & Irurtzun, A. (2021). Gravettian hand stencils as sign language formatives. *Philosophical Transactions of the Royal Society of London. Series B, Biological Sciences, 376*(1824), 20200205.

Hayden, B. (2021). Keeping count: On interpreting record keeping in prehistory. *Journal of Anthropological Archaeology, 63*, 101304.

Leroi-Gourhan, A. (1967). Les mains de Gargas: Essai pour une étude d'ensemble [The hands of Gargas: An essay for a general study]. *Bulletin de la Société Préhistorique Française, 64*(1), 107–122.

Lin, J., Wu, Y., & Huang, T. S. (2000). Modeling the constraints of human hand motion. In *Proceedings from HUMO '00: Workshop on human motion* (pp. 121–126). Institute of Electrical and Electronics Engineers (IEEE) Computer Society.

Luquet, G. H. (1926). *L'art et la religion des hommes fossils [The art and religion of fossil men]*. Masson et Cie.

Luquet, G. H. (1938). Sur les mutilations digitales [On digital mutilation]. *Journal de Psychologie, 35*, 548–598.

McCauley, B., Maxwell, D., & Collard, M. (2018). A cross-cultural perspective on Upper Palaeolithic hand images with missing phalanges. *Journal of Paleolithic Archaeology, 1*(4), 314–333.

Overmann, K. A. (2013). Material scaffolds in numbers and time. *Cambridge Archaeological Journal, 23*(1), 19–39.

Overmann, K. A. (2014). Finger-counting in the Upper Palaeolithic. *Rock Art Research, 31*(1), 63–80.

Overmann, K. A. (2021). Finger-counting and numerical structure. *Frontiers in Psychology, 12*, 723492.

[18] Verhoeven (2024).

Overmann, K. A. (2023). *The materiality of numbers: Emergence and elaboration from prehistory to present*. Cambridge University Press.

Owens, D., & Hayden, B. (1997). Prehistoric rites of passage: A comparative study of transegalitarian hunter–gatherers. *Journal of Anthropological Archaeology, 16*(2), 121–161.

Pike, A. W. G., Hoffmann, D. L., García-Diez, M., et al. (2012). U-series dating of Paleolithic art in 11 caves in Spain. *Science, 336*(6087), 1409–1413.

Rouillon, A. (2006). Au Gravettien, dans la grotte Cosquer (Marseille, Bouches-du-Rhône), l'Homme a-t-il compté sur ses doigts? [In the Gravettian, in the Cosquer Cave (Marseille, Bouches-du-Rhône), did man count on his fingers?]. *L'Anthropologie, 110*(4), 500–509.

Snow, D. R. (2013). Sexual dimorphism in European Upper Paleolithic cave art. *American Antiquity, 78*(4), 746–761.

Verhoeven, M. (2024). Between and beyond: The role of caves in Upper Palaeolithic art and cognition. In T. Wynn, K. A. Overmann, & F. L. Coolidge (Eds.), *The Oxford handbook of cognitive archaeology* (pp. 1043–1064). Oxford University Press.

Chapter 15
The Taï Plaque

Abstract The Taï plaque is a piece of bovine rib from the Grotte du Taï, a 12,000-year-old site in France. Though only 3 inches long, the bone fragment was inscribed with more than a thousand marks. As might be surmised from the minute size of the artifact, the marks it contains are so small they can be seen clearly only under magnification. The marks are also not grouped in any way detectable by unassisted visual inspection. Scholars like Alexander Marshack have speculated that the marks were accumulated over a period of several years, and that the artifact perhaps functioned as a lunar calendar. However, the miniscule size of the marks, when coupled with their large quantity, lack of overt grouping, and the fixed nature of inscriptions, make the marks difficult to see and then differentiate for the purpose of counting them. Their large quantity would also make it difficult to keep track of their number with human memory, especially over the several years involved in their accumulation. Nonetheless, because one or more individuals may have been dedicated to the task of accumulating the marks over a period of several years, the Taï plaque suggests the society that made the artifact was specialized and complex.

Keywords Specialization · Social complexity · Visual discrimination · Counting · Number sense

The Taï plaque (Fig. 15.1) is a piece of bovine rib from the Grotte du Taï, a 12,000-year-old site in the southeast of France. Despite its miniscule size—its length is about the same as an adult index finger—it's inscribed with more than a thousand marks. Each of those marks is just 0.04–0.08 inches in length, which is so small the marks can be seen clearly only under magnification. The marks may have been made over a period of several years,[1] and they are not noticeably grouped. Because it's possible to make a daily mark without counting the total produced or even being able to count, the Taï marks don't necessarily mean that their prehistoric makers had numbers, even if they look to us like they could have had numerical

[1] Marshack (1973, 1991a, 1991b).

K. A. Overmann, *Cultural Number Systems*, Interdisciplinary Contributions to Archaeology, https://doi.org/10.1007/978-3-031-83383-0_15

Fig. 15.1 Two views of an engraved bovine rib fragment from the Grotte du Taï, dated to 12,000 years ago. The artifact measures 3.4 inches long and 1.1 inches tall. The marks, over a thousand in all, are just 0.04–0.08 inches in length. Adapted and relabeled from Marshack (1991b, Fig. 1, p. 26) and used with the permission of Cambridge University Press

meaning. Accumulation without overt grouping, especially over a period of several years, suggests that making the marks was important, but their later readability, intelligibility, and reuse as a specific quantity were not. This leaves uncertain both the purpose of the marks and their potential for numerical meaning.

Alexander Marshack thought the device was a lunar calendar. Marshack, an influential archaeologist who specialized in interpreting inscribed prehistoric artifacts, said that under the microscope, the Taï marks could be seen as structured in groups of 29.[2] Since the moon takes 29.5 days to go from new to full and back again, Marshack suggested that each mark recorded a single night's lunar observation. If this were true, it would mean that the marks were accumulated over a 3½-year period. Marshack said that the marks on the reverse side of the artifact possibly indicated a "fourth ritual or economic year."[3]

It's important to emphasize that the marks are not grouped in a way that can be seen without a microscope. If Marshack was correct in his interpretation of the device, not even the number of months it represented could be easily determined by counting, since any grouping isn't readily apparent. The lack of any overt grouping,

[2] Marshack (1973, 1991b).

[3] Marshack (1991a), p. 86.

in turn, suggests that what was important about the marks was their accumulation, rather than their total as either individuals or nascent groups, and this in turn implies that their meaning was neither numerical nor calendrical per se. The Taï marks might nonetheless be consistent with an apprenticeship or esoteric purpose, since the simple fact of their accumulation would attest to the completion of observational or ritual activity.

Large quantities of undifferentiated marks imply a number system that is not highly elaborated. Anthropologist Brian Hayden has argued that the quantity of marks on the plaque is "difficult to conceive of as existing without complex numbering systems of comparable magnitudes."[4] On the other hand, a complex system of numbers carries with it certain expectations about how numbers should behave, and these expectations decrease the usability of undifferentiated marks, especially in large quantities. For example, consider Western numbers, which are highly elaborated. People enculturated into the Western number system expect numerical representations to be accessible, since the Western notations they typically use consolidate and concisely represent the units, tens, hundreds, thousands, and so on. Since the marks on the Taï plaque are not overtly grouped, they are neither consolidated nor concise, and so we would not find them particularly accessible as numerical information. Our expectations interfere with our ability to access the information as numbers in the way it's presented on the artifact. But the lack of accessibility is a problem only when accessibility is expected. If there's no such expectation, the constraint disappears.

A society like the Californian Pomo (Chap. 33) would also likely find the Taï marks difficult to use. The Pomo method of counting and recording numbers with beads and sticks involved replacing quantities of smaller units, typically strings of beads, with single markers that meant a larger unit, typically a stick of a certain length.[5] This strategy made the Pomo method of recording numbers highly concise, which in turn increased the accessibility of their numerical information. The need to count frequently, especially to higher numbers, influences counting toward becoming more efficient and effective, and bundling is a common strategy for gaining greater efficiency and effectiveness. Simply, by the time a society is counting into the hundreds, let alone the thousands, numerical representations are unlikely to have remained unbundled or ungrouped—it's just too inefficient and unwieldy. Beyond its negative effects on the accessibility of the information, the lack of bundling in the Taï marks, in and of itself, suggests that the associated number system was unlikely to have been highly elaborated.

Our perceptual system for quantity can't appreciate the Taï marks numerically. Our innate ability to perceive quantity lets us rapidly and unambiguously appreciate and identify *one*, *two*, and maybe *three* items, an ability known as *subitization*. Things start to get fuzzy around *three or four* items, the upper end of the subitizing range, which is why it is not uncommon for restricted number systems to have an

---

[4] Hayden (2021), p. 16.

[5] Barrett (1952); Loeb (1926).

upper limit of *about three or four*.[6] Above *about four*, we see "many." We can, however, distinguish a numerically bigger "many" from a numerically smaller one, if the difference between them lies above a threshold of noticeability.[7] This is how perceptual systems work in general. For example, if we were to hold two objects or hear two sounds, qualities like weight, temperature, color, loudness, and pitch would have to differ sufficiently before we would be able to tell them apart. Because we see "many" whenever there are more than *about four* marks, the Taï marks are numerically meaningless unless they are counted.

The Taï marks are so small, they aren't easy to see or use in any quantity. As just noted, we can't appreciate a quantity of marks numbering more than *about four* just by looking at them, a limitation of our ability to perceive quantity. Visual discrimination in and of itself also becomes more difficult when things are very small. This means that if a series of marks like those on the Taï plaque were to become meaningful, it must be counted in some fashion, either as groups for a calendrical meaning (which is unlikely because any grouping the marks might have cannot be seen without a microscope) or as individuals for a numerical meaning (which is unlikely because the marks are so small, so many, and so close together).

Just think what would be involved in counting the Taï marks individually. Each mark would have to be numbered: *one*, *two*, *three*, … *five hundred and twenty*, *five hundred and twenty-one*, *five hundred and twenty-two*, … *one thousand and twelve*, *one thousand and thirteen*, *one thousand and fourteen*, and so on. Such counting assumes that a counting sequence is available, which might not have been true for the prehistoric society that made the device. The marks already counted would also need to be distinguished in some way from the marks remaining to be counted. Distinguishment is more easily accomplished if the two categories—counted and uncounted—can be physically separated from one another. This just isn't possible for the Taï marks, not only because inscriptions can't be moved once made, but also because their size makes them difficult to see in the first place.

The period of time required to accumulate the marks would strain human memory. It's possible the person(s) accumulating the marks simply remembered their total from day to day, either as individuals or as months. On the other hand, this would be difficult to do because our ability to remember information is relatively perishable and easily interrupted (readers can try this at home). This is a major reason why we use material forms like tally marks in the first place: They preserve information so we don't have to remember it. Even if the total number of marks changed infrequently rather than every month or every day, the multiyear period of time involved in their accumulation would limit their ability to be meaningful through memorization.

The Taï marks do imply specialization and social complexity. Specialization is implicit to having one or more individuals dedicate several years to accumulating the marks, and this in turn argues in favor of social complexity. As was noted for the

[6] Dehaene (2011); Piazza (2011); Piazza and Izard (2009).

[7] Dehaene (2003); Weber (1834); but also see Masin (2009).

Cosquer stencils, social complexity and wealth do not mean that the society was able to count to higher numbers, only that the conditions motivating this were present. The debate over the meaning of the Taï plaque and other inscribed prehistoric artifacts will undoubtedly continue as our analytical methods and models are further refined.

References

Barrett, S. A. (1952). *Material aspects of Pomo culture* (Vol. 20). Milwaukee Public Museum.

Dehaene, S. (2003). The neural basis of the Weber–Fechner law: A logarithmic mental number line. *Trends in Cognitive Sciences, 7*(4), 145–147.

Dehaene, S. (2011). *The number sense: How the mind creates mathematics* (Revised). Oxford University Press.

Hayden, B. (2021). Keeping count: On interpreting record keeping in prehistory. *Journal of Anthropological Archaeology, 63*, 101304.

Loeb, E. M. (1926). Pomo folkways. *University of California Publications in American Archaeology and Ethnology, 19*(2), 149–405.

Marshack, A. (1973). Analyse préliminaire d'une gravure à système de notation de la grotte du Taï (Drôme) [Preliminary analysis of a notation system engraving from the Taï Cave (Drôme)]. *Etudes Préhistoriques, 4*, 13–16.

Marshack, A. (1991a). *The roots of civilization: The cognitive beginnings of man's first art, symbol and notation* (Revised). Moyer Bell.

Marshack, A. (1991b). The Taï plaque and calendrical notation in the Upper Palaeolithic. *Cambridge Archaeological Journal, 1*(1), 25–61.

Masin, S. C. (2009). The (Weber's) law that never was. In M. A. Elliott, S. Antonijević, S. Berthaud, P. Mulcahy, B. Bargary, C. Martyn, & H. Schmidt (Eds.), *Proceedings of the 25th annual meeting of the International Society for Psychophysics* (pp. 441–446). International Society for Psychophysics.

Piazza, M. (2011). Neurocognitive start-up tools for symbolic number representations. In S. Dehaene & E. Brannon (Eds.), *Space, time and number in the brain: Searching for the foundations of mathematical thought* (pp. 267–285). Academic.

Piazza, M., & Izard, V. (2009). How humans count: Numerosity and the parietal cortex. *The Neuroscientist, 15*(3), 261–273.

Weber, E. H. (1834). *De pulsu, resorptione, auditu et tactu [On pulse, resorption, hearing and touch]*. C. F. Koehler.

Chapter 16
Non-decimal and Unusual Decimal Counting Systems

Abstract Since most of the world's number systems are decimal, prehistoric European numbers were also likely to have been decimal. However, evidence for this is scarce. Some insight comes from proto-languages, hypothetical early languages reconstructed from their modern descendants. Proto-Indo-European had a decimal system that counted into the hundreds. Some European number systems are vigesimal and may represent numbers developed before Indo-European-speaking peoples moved into Europe. One is the number system of Basque, a language spoken in northern Spain and southern France. Other vigesimal numbers are found in Celtic, Romance, and Germanic languages, as possibly spread by Viking and Saxon invaders. Duodecimal (base-12) counting is suggested by the "long hundred" and the "double gross" found in Germanic languages. Danish forms its odd decades as half-steps between even multiples of *twenty*. Breton and Welsh form their *fifty* as "half hundred," and Bova Greek builds its higher decades on *fifty*. Such formations suggest that Europe was home to several non-decimal number systems before Indo-European speaking peoples brought in their decimal numbers.

Keywords Proto-language · Writing · Vigesimal · Basque · Long hundred

What kind of numbers did prehistoric European peoples have? Decimal systems are likely, as the hand is typically the first device used for counting, and this influences counting toward grouping based on the number of fingers.[1] However, there's a lot of variability in how the hand is actually used. Some societies count on both hands, so their number systems are patterned by tens (the fingers on both hands), and possibly subgrouped by fives (the fingers on each hand). Others societies continue counting on the toes, so their number systems are patterned by twenties, and often subgrouped by fives or tens. A few societies count using the spaces between their fingers, or the finger segments with or without the thumb, and so their number systems are patterned by numbers like *four*, *six*, *eight*, *twelve*, and *fourteen*. And as we have just

[1] Overmann (2014, 2021).

K. A. Overmann, *Cultural Number Systems*, Interdisciplinary Contributions to Archaeology, https://doi.org/10.1007/978-3-031-83383-0_16

seen with the Cosquer hand stencils and Taï plaque, it's not easy to determine what a number system was like—or even if there was one—from the artifacts left behind by prehistoric peoples, even when the artifacts suggest counting.

Some insight into ancient numeracy comes from comparative linguistics. Ancestral languages are reconstructed by comparing words and their different pronunciations in descendent languages. Today, most of the languages in Europe are members of the Indo-European (IE) language family. Proto-Indo-European (PIE), IE's ancestor spoken around 4500–2500 BCE, has been reconstructed by comparing descendant languages like Hittite, Sanskrit, Greek, Albanian, and English. This reconstruction shows that PIE had a decimal system that could count into the hundreds.[2] Different hypotheses locate PIE's origin to the east of Europe, either north of the Black and Caspian Seas in the Pontic-Caspian steppe (modern Russia, Ukraine, Moldova, and Romania), or south of the Black Sea in Anatolia (modern Turkey), with IE spreading into Europe around 8000 years ago.[3] In either case, as IE speakers and their decimal system migrated westward, they would have encountered the original European inhabitants and their languages and number systems. Few, if any, traces of these original languages remain, though they undoubtedly influenced some of the pronunciation changes found across the IE language family. However, there are some hints of original counting systems, ones that are organized around numbers other than *ten* or that construct their decimal numbers in unusual ways, circumstances that evoke contacts between ancient peoples and their numbers.

Direct, albeit much later, insight comes from writing. In Europe, writing began early in the first millennium BCE with the Greeks, who borrowed and adapted a new form of writing—the alphabet, symbols that mean sounds, rather than words or syllables.[4] Within centuries, alphabetic writing had spread to the Latins. Better known as the Romans, the Latins went on to conquer much of Europe—and the rest of the world as it was then known by Europeans—by the first century CE. Not only did the Latins take writing with them, recording observations about the peoples they conquered, they also transferred it to them by teaching local governors and administrators to read and write. Christianity was another mechanism for transferring literacy, as churches, monasteries, and eventually, universities began to spread throughout Europe. Once writing was present in a particular region, it took a while longer for it to be adapted to the local languages, and then for any significant recording of those languages to occur. The earliest sources that record local languages are about a thousand years old. These fragments suggest that outside the centers of Roman influence, particularly in rural areas, vigesimal (*twenty*-based) counting was the basis for trade, particularly for agricultural products.[5]

Some European number systems are vigesimal. These systems may be quite old; their geographic correspondence with non-IE languages suggests that vigesimality

---

[2] Beekes (2011); Bomhard (2008).

[3] Bouckaert et al. (2012); Chang et al. (2015).

[4] McCarter (1974); Papadoupoulos (2016).

[5] Reichenkron (1952).

Table 16.1 Basque vigesimal decades (Rebuschi, 2013)

Nr.	Basque word	Meaning
10	*hamar*	10
20	*hogei*	20
30	*hogeitahamar*	20 + 10
40	*berrogei*	2 × 20
50	*berrogeitahamar*	40 + 10
60	*hirurogei*	3 × 20
70	*hirurogeitahamar*	60 + 10
80	*larrogei*	4 × 20
90	*larrogeitahamar*	80 + 10
100	*ehun*	100

may have originated in the Mesolithic (15,000–5000 years ago).[6] One is the number system of Basque (Euskera), a language spoken in northern Spain and southern France that is generally considered to be an *isolate*, a language with no known relatives, living or extinct. Next are several IE language. These include the Celtic languages spoken in the British Isles and France; French, a Romance language; Danish, a Germanic language; and several of the Iranian languages.[7] Only Basque is thought to be an original vigesimal system based on counting with the fingers and toes; the others appear to be instances of language contact and borrowing—and fairly recent ones, at that.[8]

Because the Basque language is considered an isolate, researchers have long speculated that the Basque people "may be direct descendants of Cro-Magnon enclaves from the [U]pper Paleolithic (38,000 to 10,000 years)."[9] While the term Cro-Magnon is now obsolete, it meant the first early modern humans to settle in Europe. Supporting evidence comes from unique DNA markers and vocabulary suggestive of Stone Age technologies; for example, the Basque word for "knife" means "cutting stone."[10] However, as the linguist Robert "Larry" Trask once noted, all modern people have Stone Age ancestors, so to conclude from Basque vocabulary that they are uniquely Stone Age remnants is "a shrieking *non sequitur*."[11] Recent genetic analyses suggest that the Basque people are descended from pre-Roman tribes, and certainly, they have inhabited a distinct and fairly isolated region for a very long time.[12] Despite these mysterious origins, the Basque system of vigesimal numbers (Table 16.1) is wholly unremarkable.

[6] Blažek (1999); Perrein (2018).

[7] Bauer (2001).

[8] Bauer (2001); Greene (1992).

[9] De Pancorbo et al. (2001), p. 224.

[10] Howard (2015).

[11] Trask (1997), p. 290.

[12] E.g., Behar et al. (2012); Günther et al. (2015); Olalde et al. (2019).

Celtic languages like Irish and Welsh inherited decimal numbers from IE and later adopted vigesimal numbers.[13] In Irish and Scots Gaelic, vigesimal numbers came and went, as they were borrowed and then replaced again by decimal ones.[14] A possible source of these borrowed numbers is suggested by the word "score," which comes from early Scandinavian. "Score," which entered the vernacular about a thousand years ago, means both a unit of *twenty* used in counting and the action of cutting a notch into a wooden tally.[15] Scandinavian vigesimal is thought perhaps to represent an independent and fairly recent emergence of *twenty*-based counting.[16]

Vigesimal counting sequences with lovely rhymes are found throughout the British Isles; the one from Wiltshire, home to the Neolithic sites Stonehenge and Avebury, counts *one* to *ten* as *ain*, *tain*, *tethera*, *methera*, *mimp*, *ayta*, *slayta*, *laura*, *dora*, *dik*.[17] These rural dialects, used in counting sheep, may be remnants of an early population that was "probably subjugated, but used as a labour force" by Saxon invaders.[18]

French, a Romance language, counts its higher decades with vigesimal formations: *seventy* is *soixante-dix*, or *sixty (and) ten*; *eighty* is *quatre-vingts*, or *four twenties*; and *ninety* is *quatre-vingt-dix*, or *four twenties (and) ten*. This seems to be a clearcut instance of borrowing, since the vigesimal formations aren't found in all French dialects; instead, standard decimal formations are used: *septante*, *huitante*, *nonante*.[19] Another sign of borrowing is their higher place in an otherwise decimal counting sequence: Recipient number systems tend to borrow higher numbers but keep their own lower ones. While the origin of these vigesimal forms is unknown, they are possibly a remnant of Gaulish, the Celtic language spoken in Gaul (today, modern France) in 58 BCE, when the Roman army first invaded the region under the command of Julius Caesar (100–44 BCE).[20] Another possibility is that vigesimal forms came from the Vikings, who raided and colonized the European coastlines between the ninth and eleventh centuries CE.[21] A Viking source would make French borrowing fairly recent, within the past thousand years.

Other European numbers are unusually formed. Peculiar numbers include the "long hundred," the *hundred* valued as 120, and the "double gross," a unit *twelve*-times its worth (1728 in decimal), both found in Germanic languages.[22] These terms

[13] Greene (1992).

[14] Bauer (2001); Gillies (2003); Greene (1975); Ni Chiosáin (1999).

[15] "Score" (2022).

[16] Eliasson (2006).

[17] Gay (1999), p. 18.

[18] Ingram (1977), p. 30; also see Gay (1999); Thomas (1939).

[19] Bauer (2001).

[20] Bauer (2001); Cæsar (1904).

[21] Reichenkron (1952); Rösler (1910).

[22] Goodare (1994); Gordon (1957); Gouldesbrough (1967); Lennard (1960); Leslie (1820); Stevenson (1889).

may represent a trace of an older duodecimal (*twelve*-based) counting system,[23] perhaps inherited from the Romans, whose fractions as represented on the Roman abacus were duodecimal rather than decimal. Or, the long hundred might have been inherited from Babylonian sexagesimal (*sixty*-based) counting, as 120 is twice 60.[24] Another peculiarity is that Danish forms its odd decades as half-steps between the even multiples of *twenty*: *sixty* is *tres* (*three twenties*), *eighty* is *firs* (*four twenties*), and *hundred* is worth *five twenties*, while *fifty* is *halvtreds* (*half-three twenties*, or [3 – ½] × 20), *seventy* is *havfjers* ([4 – ½] × 20), and *ninety* is *halvfems* ([5 – ½] × 20).[25] This isn't quite as complex as the formulas might make it appear, since it's really just a matter of subtracting *ten* (half a *twenty*) from the next higher group. Roughly half of vigesimal systems worldwide construct the odd decades by subtracting from the next higher multiple of *twenty*, which is essentially what the Danish forms do. In the Celtic languages, Breton and Welsh form their *fifty* as "half hundred."[26] Building numbers on *fifty* is found in the higher decades of Bova Greek, the Griko language thought to have descended from ancient Greek colonies in Italy and Sicily:[27] While *fifty* (*pendínta*) and *sixty* (*aššínta*) are, respectively, *five tens* and *six tens*, *seventy* (*pendínta íkosi*) is *fifty (and) twenty*; *eighty* (*pendínta tránta*) is *fifty (and) thirty*; and *ninety* (*pendínta seránta*) is *fifty (and) forty*.[28] All these formations suggest a possible influence of ancient number systems in contact with IE decimal numbers.

References

Ball, M. J., & Awbery, G. (2002). Welsh, Wales, United Kingdom. In E. Chan (Ed.), *Numeral systems of the world's languages*. Max Planck Institute. https://lingweb.eva.mpg.de/channumerals/Welsh.htm

Bauer, B. L. M. (2001). Vigesimal numerals in Romance: An Indo-European perspective. *General Linguistics, 41*(1), 21–46.

Beekes, R. S. P. (2011). The numerals. In *Comparative Indo-European linguistics: An introduction* (2nd ed., pp. 237–243). John Benjamins.

Behar, D. M., Harmant, C., Manry, J., et al. (2012). The Basque paradigm: Genetic evidence of a maternal continuity in the Franco-Cantabrian region since pre-Neolithic times. *American Journal of Human Genetics, 90*(3), 486–493.

Blažek, V. (1999). *Numerals: Comparative etymological analyses and their implications: Saharan, Nubian, Egyptian, Berber, Kartvelian, Uralic, Altaic and Indo-European languages.* Masarykova Univerzita.

[23] Ulff-Møller (2023); Von Mengden (2006).

[24] Bauer (2001).

[25] Eliasson (2006); Grønnum (2002); Menninger (1992).

[26] Ball and Awbery (2002); Denis (2002).

[27] Rohlfs (1950).

[28] Reichenkron (1952).

Bomhard, A. R. (2008). Some thoughts on the Proto-Indo-European cardinal numbers. In J. D. Bengtson (Ed.), *In hot pursuit of language in prehistory. Essays in the four fields of anthropology in honor of Harold Crane Fleming* (pp. 213–222). John Benjamins.

Bouckaert, R., Lemey, P., Dunn, M., et al. (2012). Mapping the origins and expansion of the Indo-European language family. *Science, 337*(6097), 957–960.

Cæsar, J. (1904). *Cæsar's commentaries on the Gallic War* (W. A. McDevitte & W.S. Bohn, Trans.). Lee & Shepard.

Chang, W., Hall, D., Cathcart, C., & Garrett, A. (2015). Ancestry-constrained phylogenetic analysis supports the Indo-European steppe hypothesis. *Language, 91*(1), 194–244.

De Pancorbo, M. M., López-Martínez, M., Martínez-Bouzas, C., et al. (2001). The Basques according to polymorphic Alu insertions. *Human Genetics, 109*(2), 224–233.

Denis, G. (2002). Breton, Brittany region, France. In E. Chan (Ed.), *Numeral systems of the world's languages*. Max Planck Institute. https://lingweb.eva.mpg.de/channumerals/Breton.htm

Eliasson, S. (2006). Old Danish vigesimal counting: A comparison with Basque. In A. Hornscheidt, K. Kotcheva, T. Milosch, & M. Rießler (Eds.), *Grenzgänger. Festschrift zum 65. Geburtstag von Jurij Kusmenko [Border crossers. Festschrift for the 65th birthday of Jurij Kusmenko]* (pp. 91–110). Nordeuropa-Institut der Humboldt-Universität.

Gay, T. (1999). Rural dialects and surviving Britons. *British Archaeology, 46*, 18.

Gillies, W. (2003). Scottish Gaelic, United Kingdom. In E. Chan (Ed.), *Numeral systems of the world's languages*. Max Planck Institute. https://lingweb.eva.mpg.de/channumerals/ScotsGaelic.htm

Goodare, J. (1994). The long hundred in medieval and early modern Scotland. *Proceedings of the Society of Antiquaries of Scotland, 123*, 395–418.

Gordon, E. V. (1957). *An introduction to Old Norse* (2nd rev. ed.). Clarendon Press.

Gouldesbrough, P. (1967). The long hundred in the Exchequer rolls. *Scottish Historical Review, 46*(141), 79–82.

Greene, D. (1975). The Irish numerals of Cardiganshire. *Studia Celtica, 10*, 305–311.

Greene, D. (1992). Celtic. In J. Gvozdanović (Ed.), *Indo-European numerals, Part 1* (pp. 497–554). Mouton de Gruyter.

Grønnum, N. (2002). Danish, Denmark, Greenland. In E. Chan (Ed.), *Numeral systems of the world's languages*. Max Planck Institute. https://lingweb.eva.mpg.de/channumerals/Danish.htm

Günther, T., Valdiosera, C., Malmström, H., et al. (2015). Ancient genomes link early farmers from Atapuerca in Spain to modern-day Basques. *Proceedings of the National Academy of Sciences of the United States of America, 112*(38), 11917–11922.

Howard, G. (2015). The Stone-Age roots of the Basque language. *Pyrenean Experience: Basque Culture and Tradition*. https://www.pyreneanexperience.com/basque-language-origins/

Ingram, A. (1977). *Shepherding tools and customs*. Shire Publications.

Lennard, R. (1960). The long and short hundred in agrarian statistics. *Agricultural History Review, 8*(2), 99–102.

Leslie, J. (1820). *The philosophy of arithmetic; exhibiting a progressive view of the theory and practice of calculation, with an enlarged table of the products of numbers as far as one thousand* (2nd ed.). William and Charles Tait.

Marshack, A. (1973). Analyse préliminaire d'une gravure à système de notation de la grotte du Taï (Drôme) [Preliminary analysis of a notation system engraving from the Taï Cave (Drôme)]. *Etudes Préhistoriques, 4*, 13–16.

McCarter, P. K. (1974). The early diffusion of the alphabet. *Biblical Archaeologist, 37*(3), 54–68.

Menninger, K. (1992). *Number words and number symbols: A cultural history of numbers* (P. Broneer, Trans.). Dover Publications.

Ni Chiosáin, M. (1999). Irish Gaelic, Ireland. In E. Chan (Ed.), *Numeral systems of the world's languages*. Max Planck Institute. https://lingweb.eva.mpg.de/channumerals/Irish.htm

Olalde, I., Mallick, S., Patterson, N., et al. (2019). The genomic history of the Iberian Peninsula over the past 8000 years. *Science, 363*(6432), 1230–1234.

Overmann, K. A. (2014). Finger-counting in the Upper Palaeolithic. *Rock Art Research, 31*(1), 63–80.

Overmann, K. A. (2021). Finger-counting and numerical structure. *Frontiers in Psychology, 12*, 723492.

Papadoupoulos, J. K. (2016). The early history of the Greek alphabet: New evidence from Eretria and Methone. *Antiquity, 90*(353), 1238–1254.

Perrein, C. (2018). La numération vigésimale dans l'écozone paléarctique est-elle d'origine mésolithique? [Is vigesimal counting in the Palearctic ecozone of Mesolithic origin?]. *Anthropos, 113*, 467–489.

Rebuschi, G. (2013). Basque, País Vasco, Spain, France. In E. Chan (Ed.), *Numeral systems of the world's languages*. Max Planck Institute. https://lingweb.eva.mpg.de/channumerals/Basque.htm

Reichenkron, G. (1952). Einige grundsätzliche Bemerkungen zum vigesimalsystem [Some basic remarks on the vigesimal system]. In *Festgabe Ernst Gamillscheg zu seinem fünfundsechzigsten Geburtstag am 28. Oktober 1952 [Ernst Gamillscheg's commemorative publication on his sixty-fifth birthday on October 28, 1952]* (pp. 164–185). Max Niemeyer.

Rohlfs, G. (1950). *Historische grammatik der unteritalienischen Gräzität [Historical grammar of Lower Italian Greek]*. Akademie der Wissenschaften.

Rösler, M. (1910). Das Vigesimalsystem im Romanischen [The vigesimal system in Romanesque]. *Beihefte zur Zeitschrift für Romanische Philologie, 26*, 187–205.

Score, N. V. (2022, December). In *Oxford English Dictionary Online*. Oxford University Press. https://www-oed-com.libproxy.uccs.edu/view/Entry/173033

Stevenson, W. H. (1889). The long hundred and its use in England. *Archaeological Review, 4*(5), 313–327.

Thomas, D. G. M. (1939). *Animal call-words: A study of human migration*. W. Spurrell & Son.

Trask, R. L. (1997). *The history of Basque*. Routledge.

Ulff-Møller, J. (2023). The "practica" arithmetic in Icelandic-Scottish "long hundred" calculation. In A. Ries-Bund & R. Gebhardt (Eds.), *Rechenkunst und Mathematik in der frühen Neuzeit – Wissenschaftlices Kolloquium [Calculus and mathematics in the early modern period – Scientific colloquium]* (pp. 1–20). Author.

Von Mengden, F. (2006). How myths persist: Jacob Grimm, the long hundred and duodecimal counting. In G. Knappe (Ed.), *Englische Sprachwissenschaft und Mediävistik: Standpunkte – Perspektiven – Neue Wege [English linguistics and Medieval studies: Viewpoints – perspectives – new directions]* (pp. 201–209). Peter Lang.

Chapter 17
Greek Numbers

Abstract The ancient Greeks inherited many of their numerical concepts and methods from the Babylonians and Egyptians. They also had their own peculiar notions about numbers. They thought that numbers were somehow real, the Platonic realism that many mathematicians continue to accept today. They also didn't think that *one*, *two*, or *zero* were numbers, for reasons having to do with metaphysical philosophy: *one* was the *unity*, *two* was the *dyad*, and *zero* represented *nonbeing*, notions that wouldn't change until fairly recent times. The ancient Greek philosophers were also terrified by numbers that failed to fulfill their assigned metaphysical role; in the first millennium BCE, the mathematician Hippasus is said to have drowned himself over the irrationality of numbers like *pi* and the square root of *two*. The Greeks used several systems of numerical notations, including one comparable to the Roman numerals. They also thought numbers differed according to what they counted, so the *ten* used to count horses differed from the *ten* used to count dogs. This notion is similar to the idea that numbers are either *concrete*, connected to the things they count, or *abstract*, related only to other numbers.

Keywords Platonic realism · Universality · *Unity* · Nonbeing · *Zero*

The ancient Greeks are known for inventing philosophy, democracy, and geometry. While they made significant advances in mathematics, they inherited many numerical concepts and methods from the Babylonians and Egyptians.[1] They also had some rather peculiar notions about numbers, ones that seem very strange today.

The ancient Greeks thought that numbers were somehow real. The philosopher Plato (ca. 428–347 BCE) categorized numbers as *universals*, eternal but intangible and invisible beings that inhabited a different realm of existence than humans did. Other universals were notions like "beauty," "truth," and "justice." Plato never really explained how people managed to come into contact with these entities, let alone apprehend them. However, numbers have some undeniably peculiar properties that

[1] Joseph (2011).

K. A. Overmann, *Cultural Number Systems*, Interdisciplinary Contributions to Archaeology, https://doi.org/10.1007/978-3-031-83383-0_17

tend to support Plato's view of them. One of these properties is their *universality*. Numbers work in exactly the same way for everyone, regardless of time or place. As long as we perform the sum correctly, 2 + 2 will equal 4 and not some other number. Further, whatever we discover about numbers and regardless of when and where we discover it, we find the same things. For example, some numbers can only be divided by themselves and *one*, and these "prime" numbers are the same for all number systems, not different for each cultural system. Since we all find the same things, there is a sense in which numbers are "out there" somewhere, awaiting discovery. This universality also seems to elevate numerical and mathematical properties to the status of facts, and those facts, in turn, to truths. As the linguist Noam Chomsky once put it, "the fact that the square root of two is irrational is just like the fact that the sun has helium in it. And if we take a classical view of mathematics, then the truths of mathematics are just truths on a par with the truths of physics even."[2]

One, *two*, and *zero* weren't numbers to the ancient Greeks. The number *one* represented the *unity*, a metaphysical notion that was the source of existence itself. As recently as 1728—a mere three centuries ago!—the English encyclopedist Ephraim Chambers (1680–1740 CE) noted that the question of whether *one* was a number was still up for debate.[3] It probably didn't help that when *one* multiplied any other number, it yielded that very same number; this unique property would have reinforced the impression that *one* somehow differed from the rest. Since *one* wasn't a number, numbers began with plurality, which effectively starts with *two*. However, *two* was saddled with another metaphysical notion, the *dyad* or "otherness." This was apparently reason enough for the mathematician Nicomachus (ca. 60–120 CE) to doubt whether *two* was really a number.[4]

As for *zero*, it wasn't a number either, for two reasons. The first involved Greek metaphysical ideas about *being* and *nonbeing*, or *existence* and *nonexistence*.[5] *Zero* was essentially *nonbeing*, a terrifying state that annihilated the very idea of existing. The other reason the Greeks didn't have a *zero* was simple: They didn't have a notation that represented the absence of any number—a material form with the potential to represent "nothing" as just another number. Such a notation wouldn't make its way into Europe for nearly two thousand years.[6]

In the Western tradition, the idea of *zero* began in Mesopotamia. About 4000 years ago, Babylonian mathematicians aligned columns of their cuneiform numbers with a blank space.[7] Some 2500 years later, Indian mathematicians filled in the blank space with a sign that meant the absence of any number,[8] not itself a number but meaning that "no number goes here!" Over the last thousand years in the West, this

---

[2] Chomsky (2004), p. 43.

[3] Chambers (1728).

[4] Evans (1977); Nicomachus (1926).

[5] Rotman (1987).

[6] Reynolds (1993); Stone (1972).

[7] Kaplan (2000); Rotman (1987).

[8] Kaplan (2000); Rotman (1987).

Table 17.1 Greek and Roman numerals (Chrisomalis, 2010, Table 4.4, p. 99 and Table 4.10, p. 110; also see Tod, 1937)

System	1	5	10	50	100	500	1000	5000	10,000	50,000	100,000
Greek	I	Γ	Δ	𐅄	H	𐅅	X	𐅆	M	𐅇	
Roman	I	V	X	L	C	D	ↀ	ↁ	ↂ	ↇ	ↈ

sign gradually acquired new meaning as a number in its own right, albeit one with unique characteristics, like its inability to divide any other number (which is why a calculator returns "error" whenever *zero* is the denominator).

Interestingly, while Roman numerals are widely believed to have been rather clumsy for calculating because they lacked a *zero*,[9] no such mystique is attached to Greek numbers, though they were very similar in structure (Table 17.1). Like the Romans, the Greeks had no problems calculating without a *zero* because they used counting boards and finger-counting systems that worked hand-in-hand with the numbers they had. We think that calculating without a *zero* must be difficult only because we have learned calculating methods that require one.

Since numbers were metaphysically important, it was critical that they behave. So-called irrational numbers like π and $\sqrt{2}$, which today we know as fractions that neither terminate nor repeat, so greatly challenged conceptions of how numbers were supposed to behave that the mathematician Hippasus (ca. 530–450 BCE) is said to have drowned. Reports of the incident differ greatly regarding what happened to him. One account has Hippasus throwing himself into the sea, driven incurably mad by the irrationality of the square root of *two*. Some say he perished—either at the hands of the gods or his fellow mathematicians—because he divulged the mere existence of irrational numbers, others because he told the secret of how they could be calculated to men who were not initiates of the philosophers' guild. Given their metaphysical importance, numbers that challenged existence itself seemed like a particularly dangerous secret to reveal, especially when put into unqualified hands. Talk about numbers being ladened with expectations!

It's worth mentioning that in 2022, π was carried out to 100 trillion digits, an effort that required nearly 158 days to compute.[10] Despite its effect on Hippasus, $\sqrt{2}$ apparently doesn't have quite the same panache, so it's only been calculated to a paltry 10 trillion digits.

Numbers differed according to what they counted. The numbers used for everyday counting and calculating were known as *arithmos*. The philosopher Aristotle (384–322 BCE) so thoroughly identified *arithmos* with the objects they counted that this became part of their *essence*, the properties essential to making numbers *numbers*. While a number like *ten* had the same "this many" regardless of what it counted, it differed in some way according to its object. So, the *ten* used with horses differed from the *ten* used with dogs,[11] even though both counted mammals.

---

[9] Chrisomalis (2020); Schlimm and Neth (2008).

[10] Iwao (2022).

[11] Klein (1992), drawing on Aristotle.

Similarly, a common if outdated view of numbers classifies them as *abstract* or *concrete*. Concrete numbers are considered to be connected to the things they count, while abstract numbers are related only to other numbers.[12] Historically, the distinction between the two was used to differentiate Western numbers, presumably abstract, from non-Western ones, which were assumed to be concrete. This now-outdated characterization emerged from nineteenth-century ideas about the way societies think; for example, Lucien Lévy-Bruhl's 1910 work *Les fonctions mentales dans les sociétés inférieures* described two modes of societal thinking, one "modern" and "scientific," the other "primitive" and "irrational."[13] Beyond such pejorative designations, the abstract–concrete distinction does not provide an accurate way to describe differences between numbers in cultural systems. Where Aristotle was concerned with numerical *essence*, today we can talk about numbers in terms of their psychological, behavioral, and material components and the properties that interact to produce their qualities and usability, and we can distinguish how and why numerical content, structure, and organization might change over time.[14]

Greek numerals were actually a lot like Roman numerals. While the symbols used from roughly 500 BCE to 50 CE differed (Table 17.1), both systems had symbols for multiples of *five* and *ten*, repeated and added these signs together to represent numbers, and neither had a *zero*. For example, in both systems, 11,234 would be represented as one 10,000 sign, one 1000 sign, two 100 signs, three 10 signs, and four 1 signs: MXHHΔΔΔIII in the Greek numerals and ↈↀCCXXXIII in the Roman ones. The Greeks had another notational system for numbers, one that represented them with the letters of the Greek alphabet. Numerical use was typically distinguished from alphabetic use by "a horizontal stroke above the numeral-phrase, but occasionally with dots placed to either side of it."[15] Structural similarities suggest that this alphabetic system may have been inspired by Egyptian demotic numerals.[16]

Numerals weren't the only way the ancient Greeks represented numbers. In order to understand the properties of numbers, the Greeks also represented them with matrixes of dots. The number *four*, for example, was represented by four dots arranged in a two-by-two square (::). *Four* was literally a square number—a square built out of *two*—with *two* being in turn the square root of *four*. These concepts and the associated terminology are still in use today, more than two thousand years later.

[12] Overmann (2018, 2021).

[13] Lévy-Bruhl (1910, 1922); see Piaget (1928) and Damerow (1996) for applications to numbers; see criticism in Chrisomalis (2005).

[14] Overmann (2019, 2023).

[15] Chrisomalis (2010), pp. 135–139.

[16] Chrisomalis (2003).

References

Chambers, E. (1728). *Cyclopædia, or, An universal dictionary of arts and sciences* (Vol. 2). J. and J. Knapton.

Chomsky, N. (2004). *The generative enterprise revisited: Discussions with Riny Huybregts, Henk van Riemsdijk, Naoki Fukui and Mihoko Zushi* (2nd ed.). Mouton de Gruyter.

Chrisomalis, S. (2003). The Egyptian origin of the Greek alphabetic numerals. *Antiquity, 77*(297), 485–496.

Chrisomalis, S. (2005). Evaluating ancient numeracy: Social versus developmental perspectives on ancient Mesopotamian numeration. *Paper Presented at the Annual Meeting of the Jean Piaget Society (Vancouver, British Columbia)*, 1–21.

Chrisomalis, S. (2010). *Numerical notation: A comparative history*. Cambridge University Press.

Chrisomalis, S. (2020). *Reckonings: Numerals, cognition, and history*. MIT Press.

Damerow, P. (1996). Prehistory and cognitive development. In M. Killen (Ed.), *Invited lecture at the twenty-fifth annual symposium of the Jean Piaget Society Berkeley, June 1 – June 3, 1995* (pp. 1–37). Max Planck Institute for the History of Science.

Evans, G. R. (1977). From abacus to algorism: Theory and practice in medieval arithmetic. *British Journal for the History of Science, 10*(2), 114–131.

Iwao, E. H. (2022). Even more pi in the sky: Calculating 100 trillion digits of pi on Google Cloud. *Google Cloud*. https://cloud.google.com/blog/products/compute/calculating-100-trillion-digits-of-pi-on-google-cloud

Joseph, G. G. (2011). *The crest of the peacock: Non-European roots of mathematics* (3rd ed.). Princeton University Press.

Kaplan, R. (2000). *The nothing that is: A natural history of zero*. Oxford University Press.

Klein, J. (1992). *Greek mathematical thought and the origin of algebra*. Dover Publications.

Lévy-Bruhl, L. (1910). *Les fonctions mentales dans les sociétés inférieures [Mental functions in inferior societies]*. Librairie Félix Alcan.

Lévy-Bruhl, L. (1922). *La mentalité primitive [The primitive mentality]*. Librairie Félix Alcan.

Nicomachus. (1926). *Introduction to arithmetic* (M. L. D'Ooge, Trans.). Macmillan and Co.

Overmann, K. A. (2018). Updating the abstract–concrete distinction in ancient Near Eastern numbers. *Cuneiform Digital Library Journal, 1*, 1–22.

Overmann, K. A. (2019). *The material origin of numbers: Insights from the archaeology of the ancient Near East*. Gorgias Studies in the Ancient Near East 14. Gorgias Press.

Overmann, K. A. (2021). A new look at old numbers, and what it reveals about numeration. *Journal of Near Eastern Studies, 80*(2), 291–321.

Overmann, K. A. (2023). *The materiality of numbers: Emergence and elaboration from prehistory to present*. Cambridge University Press.

Piaget, J. (1928). Logique génétique et sociologie [Genetic logic and sociology]. *Revue Philosophique de la France et de l'Étranger, 105*, 167–205.

Reynolds, B. E. (1993). The algorists vs. the abacists: An ancient controversy on the use of calculators. *College Mathematics Journal, 24*, 218–223.

Rotman, B. (1987). *Signifying nothing: The semiotics of zero*. Stanford University Press.

Schlimm, D., & Neth, H. (2008). Modeling ancient and modern arithmetic practices: Addition and multiplication with Arabic and Roman numerals. In B. C. Love, K. McRae, & V. M. Sloutsky (Eds.), *Proceedings of the Cognitive Science Society* (Vol. 30, pp. 2097–2102). Cognitive Science Society.

Stone, W. E. (1972). Abacists versus algorists. *Journal of Accounting Research, 10*(2), 345–350.

Tod, M. N. (1937). The Greek acrophonic numerals. *Annual of the British School at Athens, 37*, 236–258.

Chapter 18
Roman Numerals

Abstract Roman numerals illustrate how numerical notations work generally. First, we don't need to know the Latin words for numbers to use the Roman numerals because numerical notations are *semasiographic*, communicating information without phonetic clues to the words involved. Second, we are likely to notice that Roman numerals are organized differently than Western numbers, and noticing such differences can help uncover the organizing principles. The Romans invented a pocket calculator, an abacus sized to fit into whatever kind of pocket might be construed for a toga. This pocket abacus could handle numbers into the millions, as well as fractions. Roman numerals and the associated abacus were designed to work together smoothly so that calculations could be carried out on the abacus and then recorded easily with the numerals. The Roman numerals and abacus were eventually replaced by numerals and algorithmic calculating methods imported from Indian and Arabic sources, a process that took centuries because of reluctance to give up the familiar devices. Roman numerals are still used more than fifteen centuries after the Roman empire collapsed because of their prestige value.

Keywords Semasiography · Pocket abacus · Prestige value · Algorithm · *Zero*

Roman numerals are familiar to anyone who has ever seen a Hollywood film, where they are used to express the year of release in a suitably august fashion. This prestige value is a primary reason why we continue to use Roman numerals more than fifteen centuries after the fall of the Roman empire (476 CE). Roman numerals illustrate some interesting aspects of how written numerals work generally.

We don't need their Latin names to use the Roman numerals. This quality is what classifies numerical notations as *semasiography*, writing that communicate information without speech; musical notation (♪♫♪) is another semasiographic form. Certainly, we don't experience numerals without speech, as we associate them with whatever they happen to be called in our own language: V is *five*, if we speak English, and not *quinque*, as the Romans had named it. This quality determines what happens when written numerals are adopted by societies and cultures that

K. A. Overmann, *Cultural Number Systems*, Interdisciplinary Contributions to Archaeology, https://doi.org/10.1007/978-3-031-83383-0_18

didn't invent them.[1] Unlike signs for speech, which must change to reflect new pronunciations and meanings, numerals aren't under the same pressure because they are semasiographic. This means that numeral forms tend to resist changing, no matter how many times they are borrowed or over how long a period of time.[2] Today, this same quality permits the global use of the Western numerals by societies with hundreds, if not thousands, of different languages.

We notice that Roman numerals are different. It's not just the difference in the signs, 1, 5, and 10 versus I, V, and X. Roman numerals lack a *zero*, as was mentioned previously. They also lack the place value that makes the Western numeral 5 worth *five* in the units place, but *fifty* in the tens place and *five hundred* in the hundreds place. In comparison, the Roman V is always worth *five*, no matter where it appears in a numeral phrase. Roman numerals are also repeated and then added, like the number 3 is formed by repeating the sign for 1 three times: III.

Probably the biggest thing we notice—or assume to be true—is that Roman numerals are difficult to use with our so-called algorithmic methods of calculation. An algorithm breaks up a complex process like long division into a series of short, simple steps: Dividing 7890 by 4, for example, starts with comparing 7 and 4 (and 4 goes into 7 one time with 3 left over). Such algorithms aren't confounded by Roman numerals just because they need a *zero*; it's also because Roman numerals don't line up in columns in quite the same way Western numerals do. They also aren't as concise, though the C for 100, D for 500, and M for 1000 are exceptions.[3] For example, 8–7, carry 1 from the tens to the units column is fairly easy to write in Western numerals, but VIII—VII, carry the X, not so much. Whatever we might think about the numerals, the Romans certainly didn't find them difficult to use because they had calculation methods designed to work with them.

The Romans invented the pocket calculator. The Roman Republic (509–27 BCE) and the Empire that followed it (27 BCE–476 CE) were vast and complex states, and their administration meant calculating with large numbers. While most Roman citizens would have used their fingers or a counting board with pebbles, the professional used a compact board with movable beads (Fig. 18.1, right), sized to fit easily into whatever kind of pocket we might construe for a toga. This pocket abacus could handle numbers into the millions, as well as fractions. Only three or four specimens have survived the ages, and most of these are made of bronze (one, now missing, was carved in ivory), creating speculation that most pocket calculators were made of wood.[4] The pocket abacus was quite small: The one in the British Museum, an eighteenth-century reproduction possibly inspired by Velserus' illustration (Fig. 18.1, left), is roughly 4½ long by 3 inches wide and only half an inch thick.[5]

[1] Overmann (2022).

[2] Overmann (2021).

[3] Chrisomalis (2010, 2020).

[4] Bruderer (2020).

[5] OA.2419 (2023).

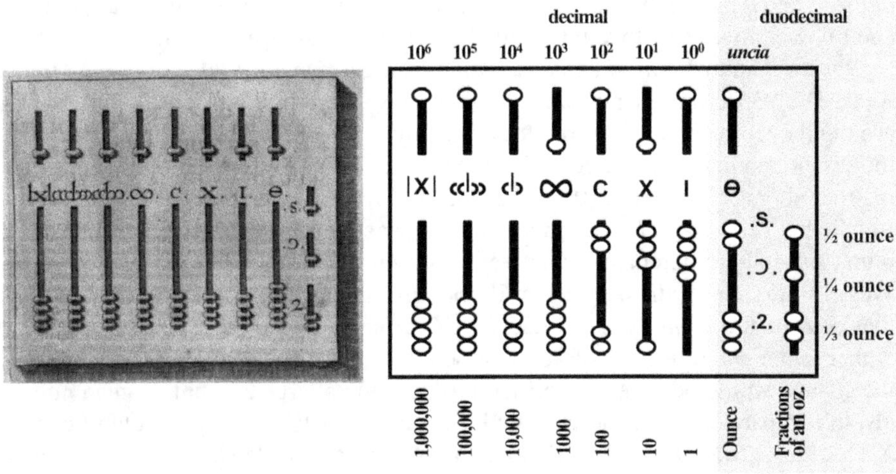

Fig. 18.1 The Roman abacus. (Left) Velserus (1682, p. 422). Image in the public domain. (Right) Image by the author based on data from multiple sources

The first seven columns of the Roman abacus represented whole decimal numbers. Column I displays the numbers 0–9 (the units), Column X 10–90 (tens), Column C 100–900 (hundreds), etc. For columns I (units) through |X| (millions), the four lower beads are valued at 1, the single upper bead at 5. In any column, 6 would be represented with one upper (5) and one lower bead (1), while exceeding 9 in any column would mean raising a lower bead (1) in the adjacent left column and returning all the beads in the exceeded column to the bottom, similar to the way a 1 would be carried in a written algorithm. Although both *zero* and place value are implicit to the abacus design—the former as all lowered beads, the latter as column value—the numbers they represented were not thought of as having either. Instead, the numerical value of each column was represented by a sign (Table 17.1) that was repeated as many times as was needed to form the desired total. So, 300 was represented by CCC, not by 3 in the hundreds place.

To the right of Column I (the units), Column ϴ represents the Roman *uncia* (ounce), or one-twelfth of the *as*, a metal coin whose weight informed its value. Fractions were used for calculating with coins, just as the other columns were, but for coins with fractional values. For the fractional columns, the decimal (base-10) system used for the other columns changes to duodecimal (base-12); hence, there are five beads in the *uncia* column, rather than four. The short column at the far right was used to calculate fractions of ounces. They are labeled .S., .Ͻ., and .2.; note that the last sign is not the Western numeral 2. Various authors have interpreted the fractions differently, so it's possible they meant something other than ½, ¼, and ⅓ ounce,[6] the values assigned by Velserus (1682), the German politician and astronomer who corresponded with Galileo regarding sunspots. On the other hand, *half*,

[6] Flegg (1983); Friedlein (1869); Ifrah (2000); Menninger (1992); Nagl (1918).

quarter, and *third* are the natural fractions formed by dividing things in half, fourths, and thirds, amounts that would be quite useful in market transactions.

Roman numerals and the abacus dominated Europe until the end of the Middle Ages. The Western numerals we use today (0–9) were first introduced into Europe around the tenth century CE, but they failed to catch on. In the twelfth century CE, the Pisano mathematician Leonardo Bonacci (1170–1250 CE), famous for originating the Fibonacci sequence (0, 1, 1, 2, 3, 5, 8, 13, 21, 34,…), (re)introduced them to Italy in his 1202 book, *Liber abaci* (which means, *The book of calculation*).[7] At this point, the Italian merchants, who were a wealthy and powerful class, greeted the Western notations with skepticism, if not outright suspicion, because they thought that *zero* or the *cipher* made fraud much easier and thus more likely.[8] Adding a cipher to the end or middle of 28 increased its value significantly (280 or 208), and the change would be relatively difficult to detect in a large account. Roman numerals, in comparison, were more difficult for the cipher to alter, as it couldn't change XXVIII to CCLXXX or CCVIII, no matter where it was added. The Roman numerals and associated methods of calculating with them were also well-entrenched: People were very familiar with them.[9] Shifting to Western numerals involved much more than just using different symbols, since calculating with them also involved the kinds of written algorithms that we use today. As a result, the strange Western numerals and algorithmic calculation methods were resisted and their merits debated, a dialogue that persisted for several centuries.[10] While the debate raged, calculating remained a matter of moving counters on counting boards, whose surface grids were labeled to corresponded with Roman numerals.[11]

Today, a similar resistance is found in the debate over the relative merits of the QWERTY and DVORAK keyboards.[12] Most people who type learn the QWERTY layout, which was intentionally designed to be awkward and slow. Awkwardness meant typing was slower, which was necessary because the mechanical keys of early typewriters were easily tangled and jammed if typing was too fast. Electronic keyboards don't work mechanically, so they permit faster typing, and keyboard layouts can be designed for greater efficiency and speed. On the other hand, most people are functional enough with the familiar QWERTY keyboard layout, and hence, we are reluctant, both individually and societally, to adopt, relearn, and regain functionality on an alternate one.

It's worth keeping in mind that when Western explorers first ventured beyond Europe and encountered cultural number systems in the Pacific (the Philippines, Chap. 24; Polynesian, Chap. 30) and the Americas (Maya, Chap. 35; Inka, Chap.

[7] Bonacci (1202).

[8] Chrisomalis (2020); Rotman (1987).

[9] Chrisomalis (2020).

[10] Evans (1977); Pullan (1968); Reynolds (1993); Stone (1972).

[11] Murray (1930).

[12] Chrisomalis (2020).

36), the system they compared the Indigenous numbers to did not use the same Western numbers and algorithmic calculations that we use today. Rather, they used the same numerals and abacus methods that Rome once used to build their empire.

References

Bonacci, L. (1202). *Liber abaci [The book of calculation]*. http://www.hs-augsburg.de/~harsch/Chronologia/Lspost13/Fibonacci/fib_intr.html

Bruderer, H. (2020). *Milestones in analog and digital computing* (J. McMinn, Trans., Vol. 1, 3rd ed.). Springer Nature Switzerland.

Chrisomalis, S. (2010). *Numerical notation: A comparative history*. Cambridge University Press.

Chrisomalis, S. (2020). *Reckonings: Numerals, cognition, and history*. MIT Press.

Evans, G. R. (1977). From abacus to algorism: Theory and practice in medieval arithmetic. *British Journal for the History of Science, 10*(2), 114–131.

Flegg, G. (1983). *Numbers: Their history and meaning*. Dover Publications.

Friedlein, G. (1869). *Die Zahlzeichen und das elementare Rechnen der Griechen und Römer und des christlichen Abendlandes vom 7. bis 13. Jahrhundert [The numerals and elementary arithmetic of the Greeks and Romans and the Christian West from the 7th to the 13th century]*. Andreas Deichert.

Ifrah, G. (2000). *The universal history of numbers: From prehistory to the invention of the computer* (D. Bellos, E. F. Harding, I. Monk, & S. Wood, Trans.). John Wiley & Sons.

Menninger, K. (1992). *Number words and number symbols: A cultural history of numbers* (P. Broneer, Trans.). Dover Publications.

Murray, D. (1930). *Chapters in the history of bookkeeping accountancy & commercial arithmetic*. Jackson, Wylie & Co.

Nagl, A. (1918). Abacus (supplement to original 1893 article). In *Paulys Realencyclopädie der classischen Altertumswissenschaft [Pauly's real encyclopedia of classical antiquity]* (S III, pp. 5–10).

OA.2419. (2023). British Museum Collection. https://www.britishmuseum.org/collection/object/H_OA-2419

Overmann, K. A. (2021). Writing system transmission and change: A neurofunctional perspective. In G. Gabriel, K. A. Overmann, & A. Payne (Eds.), *Signs – sounds – semantics. Nature and transformation of writing systems in the ancient Near East* (pp. 93–116). Ugarit-Verlag.

Overmann, K. A. (2022). Early writing: A cognitive archaeological perspective on literacy and numeracy. *Visible Language, 56*(1), 8–44.

Pullan, J. M. (1968). *The history of the abacus*. Hutchinson & Co.

Reynolds, B. E. (1993). The algorists vs. the abacists: An ancient controversy on the use of calculators. *College Mathematics Journal, 24*, 218–223.

Rotman, B. (1987). *Signifying nothing: The semiotics of zero*. Stanford University Press.

Stone, W. E. (1972). Abacists versus algorists. *Journal of Accounting Research, 10*(2), 345–350.

Velserus, M. (1682). *Opera historica et philologica, sacra et profana [Historical and philological works, sacred and profane]*. Wolfgangi Mauritii & Filiorum.

Chapter 19
Dactylonomy (Digital Reckoning)

Abstract *Dactylonomy*, the art of expressing and calculating numbers with the fingers also known as *digital reckoning*, was common in the classical and medieval world, though today it is almost forgotten. While dactylonomy was mentioned in dozens of Latin texts and depicted on coins and other objects between the third century BCE and sixth century CE, the first detailed insight comes from manuscripts written in the seventh and eighth centuries CE but known from later (eleventh or twelfth century CE) copies. The eighth-century manuscript *De temporum ratione [The reckoning of time]* was written by Bede the Venerable, a Northumbrian monk interested in calculating the date for Easter. Bede's descriptions refer to ones from the first century CE, confirming that the dactylonomic system he knew was the same one used by classical authors. While there are no known descriptions of the actual calculations, the fingers were likely used to retain the results of interim calculations, rather than providing a mechanical manipulation akin to an abacus. Bede also said that dactylonomic gestures could be used to spell out alphabetic messages, a method for communicating visually, as well as for visual puns.

Keywords Algorithm · Finger-counting · Bede the Venerable · Digital reckoning · Visual pun

Dactylonomy, the art of expressing and calculating numbers with the fingers, was common in the classical and medieval world. Despite being geographically widespread and long-lived, today this ancient art is almost completely unknown. Its name comes from the Greek words *daktulos* (finger) and *nomia* (law). Also known as *digital reckoning*, dactylonomy was mentioned in dozens of Latin texts written between the third century BCE and sixth century CE, and it was depicted on coins, reliefs, stelae, mosaics, frescos, amulets, charms, and tokens (*tessurae*) from the same period.[1] Digital reckoning was practiced in Europe and the Near East

[1] Alföldi-Rosenbaum (1971); Frœhner (1884); Minaud (2006).

K. A. Overmann, *Cultural Number Systems*, Interdisciplinary Contributions to Archaeology, https://doi.org/10.1007/978-3-031-83383-0_19

throughout the Middle Ages.[2] All these depictions assumed familiarity with what was then common knowledge, so the way(s) in which the fingers were actually used in calculating does not appear to have been recorded anywhere.

Our first detailed look comes from a seventh-century manuscript, the *Romana computatio* [*Roman computing*] written in 688 CE.[3] An early manuscript on dactylonomy that is much better known is *De temporum ratione [The reckoning of time]*, written in 725 CE by the Northumbrian monk known as Bede the Venerable (ca. 673–735 CE).[4] The oldest surviving copies of Bede's treatise are dated to the eleventh or twelfth century CE. Bede reckoned time to calculate Easter,[5] which is celebrated on the first Sunday following the first full moon after the Spring equinox. Bede not only provided written and graphic descriptions of the finger positions used in dactylonomy, his characterizations of *thirty*, *sixty*, and *hundred* referred explicitly to earlier descriptions by Jerome (ca. 342–420), the theologian and later saint best known for translating the Bible into Latin. This correspondence affirms that Bede's system was the same one known by Jerome and other classical authors.[6] Bede's system is also nearly identical to descriptions in sources like the *Farhang-i Jahāngīrī*,[7] a seventeenth-century Persian dictionary, suggesting the system possibly reentered Europe by way of the Near East, where scholars critically preserved Western knowledge after the collapse of the Roman Empire in 476 CE.[8] Interestingly, Easter wasn't calculated with dactylonomy. Instead, Bede kept track of the lunar months, solar cycles, and leap periods in the Julian calendar on his knuckles.[9] This knuckle mnemonic is similar to the way we might use these hand features today to recall which months have 31 days in them: starting from the outside edge of one hand, knuckles represent the 7 months with 31 days (January, March, May, July, continue on the second hand, August, October, and December), while the shallows between them represents the 5 months with 30 or fewer days (February, April, June, continue on the second hand, September, and November).

Dactylonomy was practiced in Europe, Africa, and the Near East for more than a thousand years. Several of the Latin texts that mention it were written by authors from Numidia, the Roman-held African territory that is today the northernmost portion of Algeria, Libya, Morocco, and Tunisia. Bede was from Northumbria, the Anglo-Saxon kingdom that spanned northern England and southeastern Scotland and one of the most distant territories ever conquered by Rome. Dactylonomy was recorded in Arabia, where the hands were reversed from the positions documented

---

[2] Al-Uqlīdisī (952); Pacioli (1494).

[3] Jones (1939); Turner (1951); Williams and Williams (1995).

[4] Wallis (1999).

[5] Fernandes (2021).

[6] Bechtel (1909).

[7] De Sacy (1823); Guyard (1871); Lemoine (1932).

[8] Gul-Chīn (1818).

[9] Bede (1100); Lemoine (1932).

by Bede[10]; this was consistent with the direction numerals were written, the highest to the lowest value from right to left.[11] Dactylonomy was also practiced in Persia—the same Achaemenid Empire of Xerxes I and his unusual method of counting his army (Chap. 6)—as early as the first century CE.[12] Eight hundred years later, the Persian mathematician Abū al-Wafā Būzhjānī (940–998 CE) promoted dactylonomy in his popular treatises for craftsmen, businessmen, and scribes.[13] The Pisano mathematician Bonacci mentioned dactylonomy as a technique for calculating with the Hindu–Arabic numerals used in his 1202 book, *Liber abaci*.[14] Dactylonomy was likely preferred as a means of calculating because it required just the hand, since other methods—writing materials (and the literacy needed to use them), abacuses, and counting boards—were expensive and rare for much of history.[15]

Dactylonomy could handle numbers into the millions. Digital reckoning didn't simply use the fingers to count from *one* to *ten*, like we might do today. Instead, the fingers of both hands, as well as various positions of the hands in relation to the body, expressed numbers up to the millions, though for most day-to-day purposes, numbers up to the hundreds likely sufficed. According to Bede (Fig. 19.1), the units were displayed with the left little, ring, and middle fingers, the tens by the left index finger and thumb. The corresponding fingers of the right hand kept track of, respectively, the thousands and hundreds. Tens of thousands involved touching various parts of the body with left hand, hundreds of thousands by the same positions with the right hand. For example, 10,000 was displayed by placing the left hand flat in the middle of chest with fingers raised toward the neck, 60,000 by grasping the left thigh with the flattened hand, and 90,000 by grasping the hip with the thumb turned toward the groin. A million was signified by clasping the hands together with interlocked fingers. Of course, these higher numbers were necessarily rounded because it isn't possible to form the smaller digits with fingers that are interlocked or grasping the thigh.

Any calculation methods, however, were never recorded. While modern scholars have offered plausible reconstructions based on Bede's system,[16] Bede himself never explained how to perform even simple arithmetic. Dactylonomy does seem to have been capable of supporting some rather complex calculations: The mathematical treatises of Abū al-Wafā Būzhjānī approximated square roots using dactylonomy, and his astronomical work involved trigonometric calculations for planes and spheres, also likely performed with the fingers.[17] It's also possible that the fingers, rather than providing a mechanical manipulation akin to an abacus, were used only

---

[10] Pellat (1977).

[11] Chrisomalis (2010).

[12] Plutarch (1961).

[13] Bloom (2001).

[14] Bonacci (1857); also see Friedlein (1869).

[15] Smith (1925).

[16] E.g., Menninger (1992); Richardson (1916); Wallis (1999).

[17] Bloom (2001); Moussa (2011).

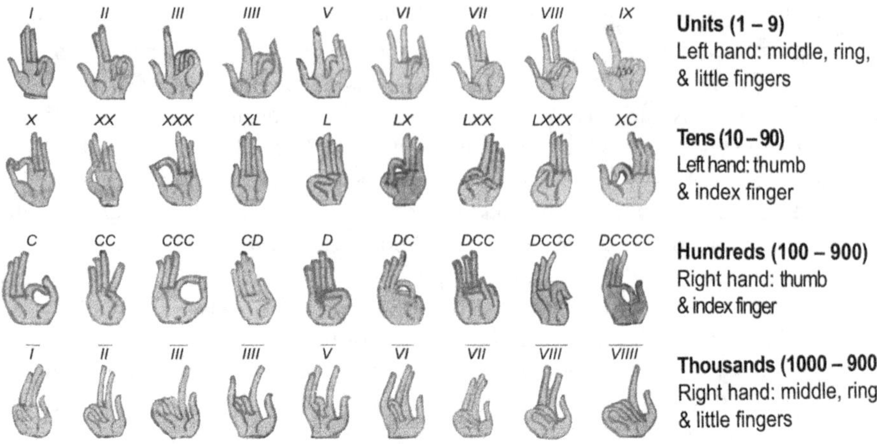

Fig. 19.1 Bede's system of digital reckoning. The line above the Roman numerals in the bottom row is a vinculum; it marks the values as thousands. Bede was inconsistent in his use of back-counting, as he gives some of the Roman numerals as four repetitions of a value, rather than as one before the next higher value (e.g., IIII rather than IV). The pattern for *one* and *one thousand* is the infeasible hand configuration found in Cosquer Cave; this suggests that dactylonomy required significant practice, not just to achieve proficiency, but to be able to produce the patterns in the first place. Adapted from Bede, ca. 1100 CE. Image in the public domain

to retain interim results like "the successive sums in a series of additions" realized through a process that was primarily mental in nature.[18] Bede suggests that this was the case by noting that in division, the number being divided was said aloud, as the hands were occupied with the number dividing it.[19]

So-called peasant algorithms from the European Middle Ages also used the fingers to support mental calculation. For example, multiplying 7 × 8 involved raising two fingers on one hand and three on the other, since 7 is 5 + 2 and 8 is 5 + 3. The numbers denoted by the raised fingers were then added (2 + 3), producing the number in the tens position (50), while those denoted by the bent fingers were multiplied (3 × 2), producing the units (6), with the product being 56.[20] While such algorithms are primarily mental, the fingers do more than display the interim results.

The gestures could have other meanings. Bede mentions that the numbers could be used to communicate letters of the alphabet, with 1 standing for A, 2 for B, 3 for C, and so on. While it's tempting to view this as simply a way of circumventing the monastic rules of silence, Bede says that the gestures could be used "to warn a friend who is among conspirators" by spelling out *caute rem* ("act cautiously") with the numbers 3, 1, 20, 19, 5 and 1, 7, 5. While we might wonder why such messages would be needed often enough to memorize the system—and in religious

[18] Turner (1951), p. 69; also see Bloom (2001).

[19] Bede, ca. (1100); Wallis (1999).

[20] Smith (1925), p. 201.

company!—the gestures were surely effective only when the intended recipient and not any unfriendly parties could see and understand them. Bede thought that alphabetic use might be easier for Greeks because they used their alphabetic letters as numerical notations. The intended meaning of the gestures as letters and not numbers would presumably come from context, as Bede doesn't mention how a vinculum would be expressed.

Dactylonomy was also used for visual puns. A second-century riddle—"take *seven* from *eight*, and *six* will remain"—reflected differences in the fingers used to make these numbers.[21] In Persia, the closed fist was synonymous with avarice, so the number *ninety-three*—also a closed fist—denoted stinginess.[22] As described by Bede, the number *thirty* was connected with marriage, "for the very joining of the fingers, as if embracing and allying each other with a tender kiss, depicts the husband and wife."[23] *Sixty* represented widowhood and the loss of sexual comfort; *one hundred*, virginity and an intact hymen.[24] Keep in mind the Romans named the middle finger *digitus impudicus*—the shameless finger—and the Greeks used the same gesture for centuries before the Romans did,[25] so the use of finger gestures to convey certain anatomical suggestions has been around for a very long time.

References

Alföldi-Rosenbaum, E. (1971). The finger calculus in antiquity and in the Middle Ages: Studies on Roman game counters I. *Frühmittelalterliche Studien, 5*(1), 1–9.

Al-Uqlīdisī, A. (952). *The arithmetic of Al-Uqlīdisī: The story of Hindu-Arabic arithmetic as told in Kitāb al-fuṣūl fī al-Ḥisāb al-Hindī* (A. S. Saidan, Trans.). D. Reidel Publishing.

Bechtel, E. A. (1909). Finger-counting among the Romans in the fourth century. *Classical Philology, 4*(1), 25–31.

Bede. (1100). Royal MS 13 A XI. Miscellany of works on computus and astronomy. *Digitised manuscripts*. https://www.bl.uk/manuscripts/FullDisplay.aspx?index=0&ref=Royal_MS_13_A_XI

Bloom, J. M. (2001). *Paper before print: The history and impact of paper in the Islamic world*. Yale University Press.

Bonacci, L. (1857). *Il liber abbaci di Leonardo Pisano [The book of the abacus by Leonardo Pisano]*. Baldassarre Boncompagni.

Chrisomalis, S. (2010). *Numerical notation: A comparative history*. Cambridge University Press.

Conley, T. (2010). *Toward a rhetoric of insult*. University of Chicago Press.

De Sacy, S. (1823). De la manière de compter, Au moyens des jointures des doigts, usitée dans l'Orient [On the manner of counting, by means of the joints of the fingers, used in the East]. *The Asian Journal, 3*, 65–71.

Fernandes, M. J. (2021). *Finger-counting in two illuminated grammatical manuscripts (12th–13th centuries)*. Unpublished master's thesis, Universidade Nova de Lisboa.

---

21 Sanford (1928).

22 Bloom (2001).

23 Sanford (1928), p. 590.

24 Wallis (1999).

25 Conley (2010).

Friedlein, G. (1869). *Die Zahlzeichen und das elementare Rechnen der Griechen und Römer und des christlichen Abendlandes vom 7. bis 13. Jahrhundert [The numerals and elementary arithmetic of the Greeks and Romans and the Christian West from the 7th to the 13th century]*. Andreas Deichert.

Frœhner, W. (1884). *Le comput digital [The digital computer]*. Protat Frères.

Gul-Chīn. (1818). Persian anthology. *Journal Asiatique, VI*, 337–347.

Guyard, S. (1871). Chapitre de la préface du Farhangi Djehangiri, sur la dactylonomie [Chapter of the preface of Farhangi Djehangiri, on dactylonomy]. *Journal Asiatique, XVIII*, 106–124.

Jones, C. W. (1939). *Bedae pseudepigraphia: Scientific writings falsely attributed to Bede*. Cornell University Press.

Lemoine, J.-G. (1932). *Revue des études islamiques. Les anciens procédés de calcul sur les doigts en orient et en occident [Review of Islamic studies. Ancient finger calculation methods in the East and West]*. Librairie Orientaliste Paul Geuthner.

Menninger, K. (1992). *Number words and number symbols: A cultural history of numbers* (P. Broneer, Trans.). Dover Publications.

Minaud, G. (2006). Des doigts pour le dire. Le comput digital et ses symboles dans l'iconographie romaine [Fingers to tell it. Digital computing and its symbols in Roman iconography]. *Histoire & Mesure, 21*(1), 3–34.

Moussa, A. (2011). Mathematical methods in Abū al-Wafāʾ's Almagest and the Qibla determinations. *Arabic Sciences and Philosophy, 21*, 1–56.

Pacioli, L. (1494). *Summa de arithmetica, geometria, proportioni et proportionalita [Summary of arithmetic, geometry, proportion and proportionality]*. Paganini.

Pellat, C. (1977). *Textes arabes relatifs à la dactylonomie [Arabic texts relating to dactylonomy]*. G.-P. Maisonneuve et Larose.

Plutarch. (1961). *Plutarch's Moralia in fifteen volumes. Vol. III: 172 A–263 C* (F. C. Babbitt, Trans.). Harvard University Press.

Richardson, L. J. (1916). Digital reckoning among the ancients. *American Mathematical Monthly, 23*, 7–13.

Sanford, E. M. (1928). De loquela digitorum [On the speech of the fingers]. *The Classical Journal, 23*(8), 588–593.

Smith, D. E. (1925). *The history of mathematics. Vol. 2: Special topics of elementary mathematics*. Ginn and Company.

Turner, J. H. (1951). Roman elementary mathematics: The operations. *The Classical Journal, 47*(2), 63–74, 106–108.

Wallis, F. (1999). *De temporum ratione [On the account of the seasons]* (F. Wallis, Trans.). Liverpool University Press.

Williams, B. P., & Williams, R. S. (1995). Finger numbers in the Greco-Roman world and the early Middle Ages. *Isis, 86*, 587–608.

Part IV
Number Systems in Asia

Chapter 20
Chinese Numbers

Abstract The earliest "true" Chinese writing consists of divinatory inscriptions on scraps of bone from the Shang Dynasty (about 1200 BCE). These "oracle bones" show a writing system capable of expressing sound values and numerals counting into the tens of thousands, traits suggesting both were preceded by earlier forms. The small numerals (*one* to *four*) expressed numbers through the quantity of their elements, while the large numerals (*five* and higher) were conventions. The conventions perhaps resolved the perceptual indistinguishability of quantities above *about four*, a limitation of the number sense. Alternatively, the conventions could have been chosen as shapes that were easy to tell apart, though scholars think the forms depicted objects whose names sounded something like the words for numbers. Another hypothesis suggests the conventions depicted finger-counting, and indeed, the signs resemble the finger-counting system found in China today. However, there is no way to tell how long a gestural system like finger-counting might persist unchanged, and Neolithic signs suggest the numeral system could have emerged thousands of years before the Shang numerals.

Keywords Number sense · Finger-counting · Multiplicative organization · Shang numerals · Oracle bones

Before the 1911 revolution established the current People's Republic, China was ruled by imperial dynasties for thousands of years. The earliest and perhaps mythical, the Xia Dynasty, was founded in 2070 BCE by a legendary ruler known as Yu the Great.[1] Yu was revered for subjugating a great flood of the Yellow River, a feat hailed as "more marvelous than any ordinary miracle"[2] because of the river's massive size. Nearly 500 years later, the Xia was replaced by the Shang, the dynasty associated with the earliest "true" Chinese writing, divinatory inscriptions on scraps

[1] Ivanhoe and Van Norden (2001).

[2] Wu (1982), p. 25.

K. A. Overmann, *Cultural Number Systems*, Interdisciplinary Contributions to Archaeology, https://doi.org/10.1007/978-3-031-83383-0_20

Table 20.1 Chinese numerals (compiled from Cheng, 1983, Table 1, p. 174; Djamouri, 1994, Table 1, p. 39)

Nr.	Neolithic	Shang	Modern	Nr.	Shang	Modern	Nr.	Shang	Modern
1	|	一	一	10	|	十	100	㔾	百
2	||	二	二	20	∪	二十	1000	千	千
3	[na]	三	三	30	Ш	三十	10,000	万	万
4	[na]	亖	四	40	Ш	四十			
5	×	㐅	五	50	㐅	五十			
6	∧	亣	六	60	亣	六十			
7	+	+	七	70	⊥	七十			
8	)(	)(	八	80	)(	八十			
9	[na]	乙	九	90	[na]	九十			

Notes: Neolithic, 4000 BCE; Shang, 1200 BCE; modern, 2000 CE. [na] = [not attested]. The Shang numeral 9 was also written in the reverse direction (Cheng, 1983, Tables 1–4, pp. 174–175; Needham, 1959)

of bone dated to about 1200 BCE.[3] These "oracle bone" inscriptions not only expressed language—the quality that makes them "true writing"[4]—they are also the direct ancestors of the modern Chinese script.[5]

The oracle bones show a mature writing system.[6] This isn't surprising because Neolithic notations—some dated as early as 6600 BCE[7] and predating the earliest inscriptions of Mesopotamia and Egypt by thousands of years—suggest Chinese writing is the world's oldest. However, the graphic forms of the Neolithic notations differ enough from the Shang forms that the two are unlikely to be directly related[8] (numerals, however, may be an exception to this). The Neolithic notations and Shang inscriptions are also separated by an extensive period of time with no intermediate forms. A similar gap occurs in Mesopotamia, though proto-cuneiform notations and cuneiform writing are separated by centuries, not millennia, and across the gap, their graphic resemblances are clear. Such gaps are not uncommon, for the material record consists only of things that preserve, and in both cases, the earliest writing is found on objects and substances that are relatively perishable.

The oracle bones also show a mature number system. By the time they appeared as Shang inscriptions, written numerals (Table 20.1) expressed numbers in the tens of thousands,[9] with the highest known, 30,000, being represented by a scorpion (the sign for 10,000) with three horizontal lines (the sign for 3) superimposed on its

[3] Demattè (2010).

[4] Qiu (2000).

[5] Chalfant (1906); Qiu (2000).

[6] Qiu (2000).

[7] Li et al. (2003).

[8] Boltz (1986); Qiu (2000); also see Demattè (2010).

[9] Chrisomalis (2010); Djamouri (1994); Martzloff (1997).

tail.[10] Such large numbers are an unambiguous sign of significant age for a number system.[11] As we would expect for a mature number system, early technologies for numbers are well attested, including things like knotted ropes, counting sticks, marks inscribed on pottery[12]—and, of course, the fingers, since Chinese numbers are organized by tens. But because some of the Shang numerals resemble Neolithic notations from Banpo, a site near Xi'an dated to 4000 BCE,[13] Chinese numbers could be very old indeed.

The Shang numerals 一 *to* 亖 *expressed numbers through their quantity of elements.* This technique reflects the one-to-one correspondence of tallying, and it produces signs that unambiguously mean numbers. Quantities in this range are also directly appreciable by the perceptual system for quantity known as the "number sense."[14] In turn, direct appreciability and its upper limit of *about three or four* explain why numerals consisting of one, two, and three strokes can persist unchanged for millennia,[15] and why numerals consisting of four strokes—the upper limit of the range, where appreciability ends—eventually acquire a different form. For the same reason but with a different solution, the Romans changed their numeral IIII to IV.[16]

Above 亖, *the Shang numerals were conventions.* This was likely a strategy for dealing with the indistinguishability of quantities above *about four*. Encountering the same problem, the ancient scribes of Mesopotamia and Egypt reorganized their numerical elements into subgroups, producing characters like 𒐌 and 𓏽. In comparison, the Chinese Neolithic numerals were simple two-stroke characters,[17] perhaps meaning only shapes that were easy to tell apart. However, because Chinese writing is based on imagery, characters are expected to depict, and there are different hypotheses about the imagery employed. One hypothesis sees the numerals as depicting objects or expressing ideas whose names were roughly similar to the words for numbers.[18] While this is the technique used for the multiples of *ten* in the Egyptian system,[19] it is much less likely to be used in the range of *one* to *nine*, since the technique of representing numbers with element quantities was not just known and easier to learn and recognize, it also produces unambiguous numbers. Further, there's little agreement about what the images might have been: 四 was a mouth, yawn, breath, or rhinoceros; 五, an empty shuttle, rope with interlaced threads, meridian, or meeting; 六, a mushroom, entryway, or hut; 十, cutting or dividing;)(, separating or a knife; 九, binding, an elbow, fishhook, or worm; |, a vertical hand,

[10] Chen et al. (2020); Cook (1995).

[11] Overmann (2023).

[12] Demattè (2022).

[13] Cheng (1983), Table 1, p. 174; also see Demattè (2022).

[14] Dehaene (2011).

[15] Overmann (2021).

[16] Chrisomalis (2020).

[17] Cheng (1983).

[18] Chrisomalis (2010); Djamouri (1994); Qiu (2000).

[19] Sethe (1916).

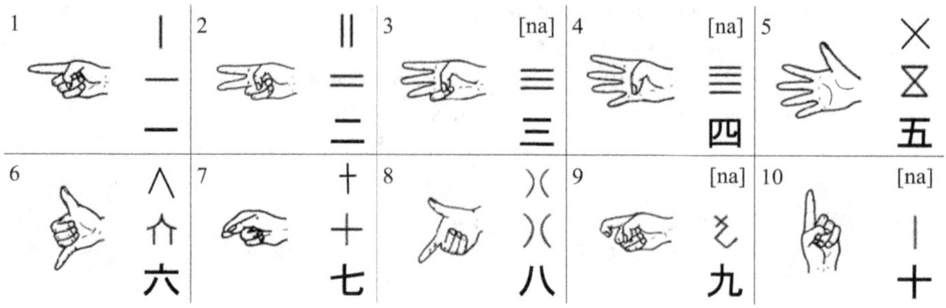

Fig. 20.1 Chinese finger-counting from the signer's perspective. The right hand shows the units, the left hand the tens. Also shown are the corresponding Neolithic (top), Shang (middle) and modern (bottom) numerals from Table 20.1. Finger-signs adapted from Domahs et al. (2010, Fig. 1, p. 253); hands and orientation from Guo (2002)

rope, or counting rod; [glyph], a thumb, inch, nose, toe, cypress cone, or container.[20] Only for [glyph] and [glyph] is there much agreement: they are generally seen as a man and a scorpion, respectively.[21] The rhino's inclusion is not nearly as silly as it might seem at first glance, as the animal was known in China during the Shang Dynasty.[22]

The Shang numerals may have depicted finger-counting.[23] Chinese finger-counting (Fig. 20.1) extends the fingers to express the numbers 1 through 4, the same logic used in the Shang numerals — through ≡. And Chinese finger-counting uses conventions for the numbers 5 through 10, just like the Shang numerals ⧖ through | do. This makes Chinese finger-counting older than the Shang numerals, which is only to be expected, since fingers are always the earliest material form for expressing numbers. But as was mentioned for Maasai finger-counting (Chap. 13), conventionality implies the finger-signs themselves were already old by the time they inspired the written forms. In fact, they could be even older, given possible links between the Shang numerals and Neolithic notations.[24] Conventionality plus the involvement of the second hand for tens suggest finger-counting that was not just extremely old but very capable, using the hands for numbers up to *ninety-nine* and then the toes for keeping track of the hundreds up to a *thousand*.[25]

As shown in Fig. 20.1, the resemblances between the finger-signs and numerals are more apparent when finger-counting begins with the right hand, held horizontally.[26] The signs — through ≡ directly correspond to the fingers used. ⧖ represents the lines on the palm; 亣 through ㄜ, the corresponding finger-patterns; |, the vertical left

[20] Djamouri (1994); Martzloff (1997); Xu (1990); Zhang (1975).

[21] Djamouri (1994).

[22] Lefeuvre (1990).

[23] Guo (2002); also see Zhang (1975).

[24] Cheng (1983), Table 1, p. 174.

[25] Martzloff (1997).

[26] Guo (2002); also see Zhang (1975).

hand; [illegible], the nose; and [illegible], the whole body.[27] On this interpretation, the modern sign 四 looks like the palm of the right hand with the folded thumb and the heart line, an interpretation that certainly supports the idea the numerals depicted the fingers.

It's possible the archaic numerals really do show finger-counting as it was practiced in ancient times. On the other hand, while finger-counting is transmitted between individuals with high fidelity in modern societies, we don't really know how long it might persist unchanged. It's also possible the modern finger-signs reflect the ancient numerals,[28] rather than being their inspiration. In this case, however, the modern finger-signs would need to have been adopted long enough ago to permit the wide prevalence observed today, as well as the absence of any older forms.

Chinese numerals are multiplicative, a rare organization. In a multiplicative system, a multiplier—a symbol meaning *ten*, *hundred*, *thousand*, etc.—follows each numeral that should be multiplied. For example, 738 is 七百三十八 in Chinese numerals: the sign for *hundred* (百) multiplies the preceding sign (七) to yield the hundreds, and the sign for ten (十) multiplies the preceding sign (三) to yield the tens. Numerals that are not followed by any multiplier (八) are not multiplied. Multiplicative numerals differ from positional notations (which encode value with place, as in Western numerals like 1, 10, and 100) because they don't include a *zero* for any omitted values; for example, 708 is 七百八, or 7 × 100 and 8 [units] with no tens. They also differ from accumulative notations (which encode value with form, as in Roman numerals like I [10], X [10], and C [100]) because the multipliers are explicitly provided as separate signs. Explicit multipliers are also used in spoken numbers: just like we would say *seven hundred and thirty-eight* in English and not *seven three eight*, the equivalent Chinese term (*qī bǎi sān shí bā*) would translate as *seven hundred three ten eight*. This language-like quality in the written organization has been used to argue that the numeral forms were based on words, rather than fingers.[29] Multiplicative numerals are interesting because the calculating methods—the abacus and counting rods—both had implicit place value.

References

Boltz, W. G. (1986). Early Chinese writing. *World Archaeology, 17*(3), 420–436.

Chalfant, F. H. (1906). *Early Chinese writing*. Carnegie Institute.

Chen, K. Y., Song, Z., Liu, Y., & Anderson, M. (2020). *Reading of Shāng inscriptions*. Springer Nature.

Cheng, T. (1983). *Studies in Chinese art*. The Chinese University.

Chrisomalis, S. (2010). *Numerical notation: A comparative history*. Cambridge University Press.

Chrisomalis, S. (2020). *Reckonings: Numerals, cognition, and history*. MIT Press.

[27] Djamouri (1994); Zhang (1975).

[28] Nishiyama (2013).

[29] Chrisomalis (2010).

Cook, R. S. (1995). The etymology of Chinese Chén. *Linguistics of the Tibeto-Burman Area, 18*(2), 1–232.

Dehaene, S. (2011). *The number sense: How the mind creates mathematics* (Revised). Oxford University Press.

Demattè, P. (2010). The origins of Chinese writing: The Neolithic evidence. *Cambridge Archaeological Journal, 20*(2), 211–228.

Demattè, P. (2022). *The origins of Chinese writing*. Oxford University Press.

Djamouri, R. (1994). L'emploi des signes numériques dans les inscriptions Shang [The use of numerical signs in Shang inscriptions]. *Extrême-Orient, Extrême-Occident, 16*(1), 12–42.

Domahs, F., Moeller, K., Huber, S., Willmes, K., & Nuerk, H.-C. (2010). Embodied numerosity: Implicit hand-based representations influence symbolic number processing across cultures. *Cognition, 116*(2), 251–266.

Guo, M. (2002). *Quanji [Complete works]. Vol. 1: Kaogu juan [On archaeology]*. Originally published in 1931 as *Jiagu wenzi yanjiu [Ancient characters research]* in Shanghai by Dadong shuju. Kexue chubanshe.

Ivanhoe, P. J., & Van Norden, B. W. (2001). *Readings in classical Chinese philosophy*. Seven Bridges Press.

Lefeuvre, J. A. (1990). Rhinoceros and wild buffaloes north of the Yellow River at the end of the Shang Dynasty. *Monumenta Serica, 39*, 131–157.

Li, X., Harbottle, G., Zhang, J., & Wang, C. (2003). The earliest writing? Sign use in the seventh millennium BC at Jiahu, Henan Province, China. *Antiquity, 77*(295), 31–44.

Martzloff, J.-C. (1997). *A history of Chinese mathematics* (S. S. Wilson, Trans.). Springer.

Needham, J. (1959). *Science and civilization in China. Vol 3: Mathematics and the sciences of the heavens and the earth*. Cambridge University Press.

Nishiyama, Y. (2013). Counting with the fingers. *International Journal of Pure and Applied Mathematics, 85*(5), 859–868.

Overmann, K. A. (2021). Writing system transmission and change: A neurofunctional perspective. In G. Gabriel, K. A. Overmann, & A. Payne (Eds.), *Signs – sounds – semantics. Nature and transformation of writing systems in the ancient Near East* (pp. 93–116). Ugarit-Verlag.

Overmann, K. A. (2023). *The materiality of numbers: Emergence and elaboration from prehistory to present*. Cambridge University Press.

Qiu, X. (2000). *Chinese writing* (G. L. Mattos & J. Norman, Trans.). The Society for the Study of Early China.

Sethe, K. (1916). *Von Zahlen und Zahlworten bei den alten Ägyptern und was für andere Völker und Sprachen daraus zu lernen ist [About numbers and number words among the ancient Egyptians and what other peoples and languages can learn from them]*. Karl J. Trübner.

Wu, K.-C. (1982). *The Chinese heritage*. Crown Publishers.

Xu, Z. (Ed.). (1990). *Jia gu wen zi dian [Oracle bone dictionary]*. Sichuan Dictionary Publishing House.

Zhang, B. (1975). Jiaguwen zhong suo jian de shu ["Number" seen in oracle bone inscriptions]. *Zhongyang Yanjiuyuan Lishi Yuyan Yanjiusuo Jikan, 46*(3), 347–390.

Chapter 21
Chinese Calculating Methods

Abstract Early Chinese calculating involved two methods: the *abacus*, a device with movable counters, and *counting rods*, short, flat sticks arranged to form numerals. Various forms of abacus are found worldwide, suggesting multiple independent inventions, and the Chinese abacus may have been imported from an external source. The original Chinese abacus was a board with loose multicolored counters; the modern version uses beads guided by wires. The counting rods were unique to China and may represent the world's earliest knowledge-based calculation. Simple arithmetic was performed on the counting rods directly because they displayed numbers with the quantity and arrangement of their elements. More complex operations involved laying down the sticks needed to represent the numbers in a calculation on different rows, leaving a space between them for the interim calculation. The earliest known counting board comes from Greece and is dated to 300 BCE; the earliest counting rods are from 100 BCE, but they are depicted on coins and pottery dated to the fourth century BCE, suggesting that they may be the older device. Both devices show signs of an even earlier origin, though the evidence is inconclusive.

Keywords Abacus · Counting rods · Counting boards · Knowledge-based calculating · Movement-based calculating

Two calculating methods were used in China: the *abacus*, a device with movable counters, and *counting rods*, sticks for writing temporary numerals. While various forms of an abacus are found worldwide, suggesting multiple independent inventions, the counting rods were a Chinese creation that may represent the world's earliest knowledge-based calculation.

An abacus works by rearranging counters. As was mentioned in conjunction with the Taï plaque (Chap. 15), it's easier to keep track of the things that have been counted if they're physically separated from the things that haven't been counted, and this is the basic idea of the abacus. Today the device consists of multiple columns of beads that slide along wires, from the (uncounted) position to the (counted) position. In comparison, ancient abacuses were loose counters moved on a flat surface: clay tokens in Mesopotamia, perhaps on surfaces with lines drawn in dust or

K. A. Overmann, *Cultural Number Systems*, Interdisciplinary Contributions to Archaeology, https://doi.org/10.1007/978-3-031-83383-0_21

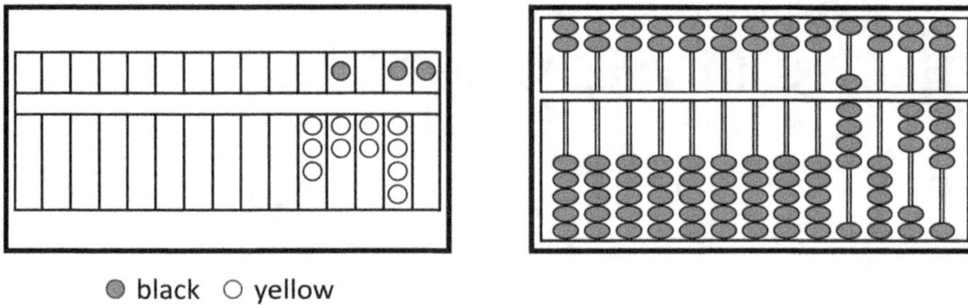

Fig. 21.1 (Left) The original Chinese abacus with loose counters, based on a description from the sixth century CE. Redrawn and relabeled from Kojima (1963, Fig. 1, p. 12). (Right) The modern Chinese abacus, called *suanpan*. Image by the author

sand; pebbles in Egypt, as reported by the Greek historian Herodotus;[1] kernels of corn or beans in the Americas, described by early Western colonizers;[2] and "jettons," coin-like objects used with counting boards in Europe, as documented in various arithmetic treatises.[3] In some cases—the European jettons, for example—the counters were identical, and their value was determined by their place within the grid. In other cases, counters differed, with shape and size (tokens, Chap. 2) or color (kernels of corn, Chap. 36) encoding their values.[4] In China, the original abacus was a counting board with loose black or yellow counters;[5] the modern version uses beads guided by wires (Fig. 21.1).

Abacuses are movement-based calculation. Using one means learning how to rearrange the counters to manipulate the numbers they represent. There are also rules for shifting counters between registers—say, from units to tens in addition or from tens to units in subtraction. For the modern abacus, the rules resemble those used with written numerals. With written numerals, we write a number like 14 as the unit (4) in the ones column and the decade (1) in the next ("carry the 1"); on an abacus, the corresponding beads in the units and tens columns are moved to the counted position, though there are a number of shortcuts unique to this method of calculation.

A modern abacus has a fixed number of beads. With a counting board, the number of counters was variable: counters were added or removed as needed for a particular calculation. When the counters in any column (or row, for horizontal grids) reached their maximum number, they would be removed and a single counter placed in the next higher column (or vice versa, in subtraction). Similar exchanges were made with Mesopotamian tokens, operations called "bundling" and "debundling." In adding, tokens were commingled[6] and then picked up and sorted, probably in the

[1] Godley (1922).

[2] De Landa (1864); Williams (1866).

[3] Barnard (1916); Murray (1930); Pullan (1968).

[4] Florio (2009); Kojima (1963); Nissen et al. (1993).

[5] Kojima (1963).

[6] Høyrup (2002).

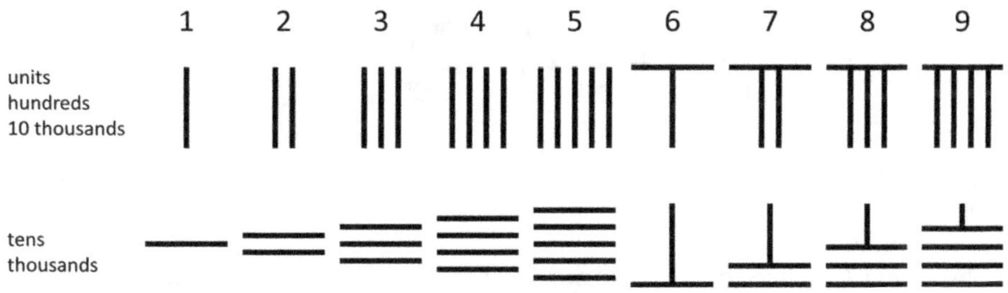

Fig. 21.2 Chinese counting rods arranged as numerals. Redrawn and relabeled from Se (1977, p. 97)

way we might organize coins as dollars today—first picking up all the coins with the largest value (quarters, in the case of the American dollar), then all the coins with smaller values (dimes, nickels, and pennies), until the total has been ascertained.

Counting rods are knowledge-based calculation. Counting rods were flat sticks, 1–6 inches long and traditionally made of bamboo,[7] that were arranged as temporary numerals (Fig. 21.2). Addition and subtraction could be performed on them directly, by adding or removing the appropriate sticks,[8] similar to the way described for the Alaskan Kaktovik numerals (Chap. 32).[9] More complex operations involved laying down the sticks needed to represent the numbers in a calculation on different rows, leaving space for an interim calculation. As the calculation proceeded from left to right (that is, from highest to smallest, the opposite of the direction typical for Western calculations), the interim number would be successively modified by rearranging the sticks displaying the interim result, with sticks added as needed as the calculation proceeded and the answer was fully realized.

As shown in the figure, adjacent powers of *ten* were distinguished by alternating the different stick arrangements. The counting rods omitted the multiplier symbols used in written numerals (e.g., 十 for tens, 百 for hundreds). This effectively made the counting rods a place value system, with blank spaces (a precursor of *zero*) representing the absence of any number.[10]

Whether used for simple visual arithmetic or complex manipulations of interim results, calculating with counting rods is the kind of knowledge-based calculation used today with written notations. In knowledge-based calculation, the initial numbers are written in some fashion, either with pencil and paper or as formed with counting rods. The large problem is then broken into a series of small steps, a procedure known as an *algorithm*. Transitioning from one step to another involves making mental judgements and recalling relations between numbers. These are written as interim products, and the final answer is produced once the algorithm has been completed. For example, in multiplying 213 × 44 with written numerals, we would start by multiplying 3 × 4, recall that 3 × 4 is 12, write this as 2 in the units

[7] Chrisomalis (2010); Cook (2022); Se (1977); Zhu (2018).

[8] Dauben (2007).

[9] Tillinghast-Raby (2023).

[10] Chrisomalis (2010); Martzloff (1997).

Fig. 21.3 (Left) The cuneiform sign saĝĝa. (Right) The Chinese sign *suan*. Image by the author. The signs are Unicode characters

column and 1 carried to the tens column, and so on. In using the counting rods[11] (as transliterated here into Western numerals), we would start by multiplying 2 × 44, recall that each 2 × 4 is 8, and use sticks to form 88 toward the left of the space for interim results; we would next multiply 1 × 44, recall that each 1 × 4 is 4, add 44 to 88(0) as assisted visually by the sticks, and rearrange the interim sticks to show 924; we would finally multiply 3 × 44, recall that each 3 × 4 is 12, add 132 to 924(0), and rearrange the interim sticks to show 9372.

The use of mental judgements and memorized relations is what makes knowledge-based calculating distinctive. While these things are also applicable to calculation as it is performed today on the modern abacus,[12] it's likely that ancient peoples simply moved their counters around as needed to add and subtract them.

We don't know how old these devices are. The earliest known physical abacus is Western, a marble counting board discovered on the Greek island of Salamis that dates to about 300 BCE.[13] The earliest known counting rods are just bit younger at about 100 BCE.[14] However, counting rods are also depicted on coins and pottery dated to the fourth century BCE,[15] suggesting they may be the older device.

Both the abacus and counting rods show signs of even earlier origins. For the abacus, one source indicating an ancient age is linguistic: the etymology of "abacus." In Hebrew, *'āḇāq* means *dust*, and in Greek, *ἄβαξ* designates a surface without a base, generally rectangular in shape.[16] The idea that lines were drawn in dust is almost certainly wrong, since they would be scuffed or erased every time the counters were moved.[17] This evidence, unfortunately, doesn't pinpoint an exact origin, just an old one. Written signs for words like "counting" or "calculating" are another indication of age, especially if they look a bit abacus-like. Examples include the cuneiform sign saĝĝa (Fig. 21.3, left), "the chief administrator of a temple household" (that is, someone who kept accounts), which is attested in Mesopotamia as early as 2500–2600 BCE.[18] The Chinese sign *suan* (Fig. 21.3, right), meaning "to calculate," is attested as early as the fourth century BCE.[19]

[11] Mikami (1913; see also Sun-Tzŭ (300 BCE).

[12] Kojima (1954, 1963).

[13] Kubitschek (1900).

[14] Se (1977); Volkov (2018).

[15] Chrisomalis (2010); Needham (1959).

[16] Kojima (1954).

[17] Woods (2017); also see Menninger (1992); Nagl (1918); Pullan (1968).

[18] Ifrah (1981); University of Pennsylvania (2021).

[19] Woods (2017); also see Martzloff (1997); Menninger (1992); Needham (1959).

Certain calculating errors provide a third source of evidence. For example, characteristic errors in lengthy Mesopotamian calculations from the Old Babylonian period (1900–1600 BCE) suggest that two abacuses were conjoined to perform them; these probably represent mistakes made when transferring counters from one abacus to another.[20]

Another evidentiary source is textual. For counting rods, the *Tao Te Ching*, the foundational text of Taoism thought perhaps to have been written as early as the sixth century BCE,[21] says that "Good mathematicians do not use counting-rods."[22] This is seen as indicating that counting rods were available but not fully accepted.[23] For the abacus, the *Shushu jiyi*—"Memoir on some traditions of mathematical art," perhaps written during the late second century CE—said it was a device that "controls all four seasons and covers the three orders—heaven, earth and men."[24] Because this description was poetical, not practical, commentary added in the mid-sixth century described a physical abacus (Fig. 21.1, left).[25] However, because the *Shushu jiyi* itself contains a number of anachronisms, it may have been fabricated,[26] and thus, it isn't a reliable source.

On the other hand, numerical grids were known in China perhaps as early as the late third millennium BCE. Around the time Yu the Great was subduing the flooded Yellow River, a small turtle emerged from one of its tributaries with a curious pattern of dots inscribed on its shell; when taken as nine numbers, this "magic square" known as the Lo Shu yields the sum 15, whether added horizontally, vertically, or diagonally.[27] While the 3 × 3 Lo Shu grid doesn't look much like an abacus or counting board, its design isn't nearly as inconsistent with the kind of layouts used with counting rods.

Whatever its initial start, by the mid-first millennium CE, counting rods were entrenched as the suitable foundation for learning and performing arithmetic,[28] perhaps preferred because they were faster, compared to the original Chinese abacus. Both methods spread to neighboring Korea and Japan.[29] Once the *suanpan* guided-bead abacus was introduced to China in the fourteenth or fifteenth century CE,[30] its faster speed made it very popular. It would ultimately displace the counting rods, to the point they had "disappeared from Chinese society"[31] by the seventeenth century CE.[32]

---

20 Woods (2017).

21 Volkov (2018).

22 Needham (1959), p. 70.

23 Chrisomalis (2010).

24 Yamazaki (1959), p. 92.

25 Ho (2008); Kojima (1963); Yamazaki (1959).

26 AuYong (2018); Volkov (2024).

27 Ho (2008); Swetz (2008).

28 Chrisomalis (2010); Dauben (2007).

29 Hoon (2006); Kojima (1954, 1963).

30 Chrisomalis (2010).

31 Hoon (2006), p. 476.

32 Kojima (1963).

References

AuYong, K.-X. (2018). *The ten mathematical classics: The mathematics and mathematical education of pre-modern China*. Unpublished doctoral thesis, University of Alberta.

Barnard, F. P. (1916). *The casting-counter and the counting-board: Chapter in the history of numismatics and early arithmetic*. Clarendon Press.

Chrisomalis, S. (2010). *Numerical notation: A comparative history*. Cambridge University Press.

Cook, C. (2022). Divination in BCE China according to newly-recovered and excavated texts: An overview. In M. Lackner & Z. Lu (Eds.), *Handbook of divination and prognostication in China. Part one: Introduction to the field* (pp. 61–122). Brill.

Dauben, J. W. (2007). Chinese mathematics. In V. Katz (Ed.), *The mathematics of Egypt, Mesopotamia, China, India, and Islam: A sourcebook* (pp. 187–384). Princeton University Press.

De Landa, D. (1864). *Relation des choses de Yucatan de Diego de Landa [Diego de Landa's account of things in Yucatan]*. Auguste Durand.

Florio, C. (2009). Encuentros y desencuentros nella individuazione di una relazione matematica nella yupana in Guaman Poma de Ayala [Encounters and despairs in the identification of a mathematical relationship in the yupana of Guaman Poma de Ayala]. In E. Guagliano (Ed.), *Incontri e disincontri tra Europa e America, Atti del XXX Convegno Internazionale di Americanistica Salerno, 14–15 maggio e 10–12 dicembre 2008 [Meetings and discussions between Europe and America, Proceedings of the XXX International Conference of American Studies Salerno, 14–15 May and 10–12 December 2008]* (pp. 151–186). Oèdipus Editore.

Godley, A. D. (1922). *Herodotus, the histories*. Harvard University Press.

Ho, P. Y. (2008). Magic squares in China. In H. Selin (Ed.), *Encyclopaedia of the history of science, technology, and medicine in non-Western cultures* (pp. 1251–1252). Springer.

Hoon, J. Y. (2006). Mathematics in context: A case in early nineteenth-century Korea. *Science in Context, 19*(4), 475–512.

Høyrup, J. (2002). *Lengths, widths, surfaces: A portrait of Old Babylonian algebra and its kin*. Springer.

Ifrah, G. (1981). *The universal history of computing: From the abacus to the quantum computer* (R. Laffont, Trans.). John Wiley & Sons.

Kojima, T. (1954). *The Japanese abacus: Its use and theory*. Charles E. Tuttle.

Kojima, T. (1963). *Advanced abacus: Japanese theory and practice*. Charles E. Tuttle.

Kubitschek, W. (1900). Die Salaminische Rechentafel [The Salaminian calculation table]. In *Numismatische Zeitschrift 1899 [Numismatic journal 1899]* (Vol. 31, pp. 394–398). Numismatiſchen Geſellſchaft.

Martzloff, J.-C. (1997). *A history of Chinese mathematics* (S. S. Wilson, Trans.). Springer-Verlag.

Menninger, K. (1992). *Number words and number symbols: A cultural history of numbers* (P. Broneer, Trans.). Dover Publications.

Mikami, Y. (1913). *The development of mathematics in China and Japan*. B. G. Teubner.

Murray, D. (1930). *Chapters in the history of bookkeeping accountancy & commercial arithmetic*. Jackson, Wylie & Co.

Nagl, A. (1918). Abacus (supplement to original 1893 article). In *Paulys Realencyclopädie der classischen Altertumswissenschaft [Pauly's real encyclopedia of classical antiquity]* (S III, pp. 5–10).

Needham, J. (1959). *Science and civilization in China. Vol 3: Mathematics and the sciences of the heavens and the earth*. Cambridge University Press.

Nissen, H. J., Damerow, P., & Englund, R. K. (1993). *Archaic bookkeeping: Early writing and techniques of economic administration in the ancient Near East* (P. Larsen, Trans.). University of Chicago Press.

Pullan, J. M. (1968). *The history of the abacus*. Hutchinson & Co.

Se, A. T. (1977). Chinese computation with the counting-rods. In *Papers on Chinese studies* (Vol. 1, pp. 97–109). University of Malaya.

Sun-Tzŭ. (300 BCE). Rod calculus: Multiplication algorithm. In Sunzi suanjing [Arithmetical classic of Sun-Tzŭ] (Conway, Trans., Vol. 1). https://yawnoc.github.io/sun-tzu/i/8

Swetz, F. J. (2008). *Legacy of the Luoshu: The 4,000 year search for the meaning of the magic square of order three* (2nd rev. ed.). A. K. Peters.

Tillinghast-Raby, A. (2023). Arctic arithmetic. *Scientific American, 328*(6), 14–16.

University of Pennsylvania. (2021). *The Pennsylvania Sumerian Dictionary Online*. Museum of Anthropology and Archaeology, University of Pennsylvania. http://psd.museum.upenn.edu/

Volkov, A. (2018). Chinese counting rods: Their history, arithmetic operations, and didactic repercussions. In A. Volkov & V. Freiman (Eds.), *Computations and computing devices in mathematics education before the advent of electronic calculators* (pp. 137–188). Springer Nature.

Volkov, A. (2024). Xu Yue [also known as: Hsü Yüeh]: Chinese astronomer and mathematician. In *Encyclopaedia Britannica*. https://www.britannica.com/biography/Xu-Yue

Williams, R. (1866). A key into the language of America: Or, an help to the language of the natives in that part of America, called New-England. In *Publications of the Narragansett Club, first series* (Vol. 1, 4th ed.). Narragansett Club.

Woods, C. (2017). The abacus in Mesopotamia: Considerations from a comparative perspective. In L. Feliu, F. Karahashi, & G. Rubio (Eds.), *The first ninety years: A Sumerian celebration in honor of Miguel Civil* (pp. 416–478). Walter de Gruyter.

Yamazaki, Y. (1959). The origin of the Chinese abacus. *Memoirs of the Research Department of the Tōyō Bunko (The Oriental Library), 18*, 91–140.

Zhu, Y. (2018). How were Western written calculations introduced into China?—An analysis of the *Tongwen suanzhi* (Arithmetic guidance in the common language, 1613). *Centaurus, 60*, 69–86.

Chapter 22
Numbers in Japan

Abstract Japan borrowed the Chinese system of writing, including its numerals and calculating methods, in the eighth century CE. While the Chinese writing was adapted to write Japanese, the borrowed numerals retained their Chinese forms and names, displacing but not replacing the earlier Native Japanese numbers. The Native Japanese numbers were formed by doubling, and scholars have speculated that the system of doubling or pairing may have been related to ancient finger-counting. However, the finger-counting found in Japan today bears no trace of pairing, though paired finger-counting is found in other parts of the world (for example, in Africa, where it takes the form of finger-counting that is symmetrical rather than sequential). Nor did the Japanese adopt the Chinese method of finger-counting, despite adopting their numerals, number-words, and calculating methods. The Ainu, an ethnic minority in northern Japan and adjacent territories, forms its distinctive vigesimal numbers by over-counting. A fully decimal version of Ainu numbers was recorded on the island of Sakhalin, now Russian territory, and may reflect contact with Manchurian occupiers during two centuries of Manchu rule.

Keywords *Kanji* · Finger-counting · Pairing · Ainu · Over-counting

In the eighth century CE, about a century after their first recorded battle,[1] Japan borrowed China's writing system, adapting the *kanji* or "Han characters" for Japanese and incorporating a fair portion of Chinese vocabulary into "Sino-Japanese."[2] The adopted *kanji* included the oracle bone numerals, attested in one of Japan's earliest books, a collection of poems known as the *Man'yōshū* (万葉集) or "Ten thousand gathered leaves."[3] Here we are interested in the earlier Native Japanese numbers, which are still used today to count small, inanimate objects, as well as the numbers

[1] Wang (2023).

[2] Frellesvig (2010).

[3] Frellesvig (2010), p. 289.

K. A. Overmann, *Cultural Number Systems*, Interdisciplinary Contributions to Archaeology, https://doi.org/10.1007/978-3-031-83383-0_22

Table 22.1 Native Japanese numbers (Frellesvig, 2010, Table 9.4, p. 289; Kurakichi, 1937, p. 1; Miller, 1969, p. 24)

Nr.	Native Japanese	Meaning	Sino-Japanese	Meaning
1	*pito*	1	*iti*	1
2	*puta*	1 + 1	*ni*	2
3	*mi*	3	*san*	3
4	*yo*	4	*si*	4
5	*itu*	5	*go*	5
6	*mu*	3 + 3	*roku*	6
7	*nana*	7	*siti*	7
8	*ya*	4 + 4	*pati*	8
9	*kokono*	3 × 3	*kiu*	9
10	*towo*	5 + 5	*zipu*	10

of the Ainu, an Indigenous people of northern Japan, eastern Russia, and adjacent islands.

Native Japanese numbers are formed by doubling. As shown in Table 22.1, doubles are the same word with altered vowels, the method used to mark plurality in English words like "goose" (the singular form) and "geese" (the plural form). The numerical pairs thus formed are *pito* ➔ *puta* (1 and 2); *mi* ➔ *mu* (3 and 6), *yo* ➔ *ya* (4 and 8), and *tu* ➔ *to* (5 and 10).[4]

Historian Kurakichi Shiratori thought the system of doubling or pairing might have been related to ancient finger-counting.[5] In modern Japan, finger-counting begins with the open hand, perhaps symbolizing a blank sheet of paper.[6] From the open hand, the fingers are folded sequentially, beginning with the thumb, so that *five* makes a fist; from the fist, the fingers are sequentially unfolded in the reverse order, returning to an open hand with *ten* counted on a single hand.[7] As Kurakichi aptly observed, these finger-signs do not readily correspond to the pairing found in the Native Japanese number-words. Nonetheless, he thought his "ancient forefathers" perhaps used both hands in counting, though it was unclear to him how they might have formed the pairs *one-two* and *two-four*.[8] He also speculated that higher pairs like *three-six* would involve extending the same number of fingers on each hand, forming two equal rows:

> Thus, over against this one row of three fingers which they called *mi*, and they had the sum of the fingers in two rows which they called *mu*. As eight and ten could be counted in a similar way, over against *yo* (four) they called eight double the number, *ya*; and over against *itu* (five), they called ten, double the number, *towo*.
>
> (Kurakichi, 1937, p. 4)

[4] Frellesvig (2010); Kurakichi (1937).

[5] Kurakichi (1937).

[6] Nishiyama (2013).

[7] Kurakichi (1937); Nishiyama (2013).

[8] Kurakichi (1937), p. 2.

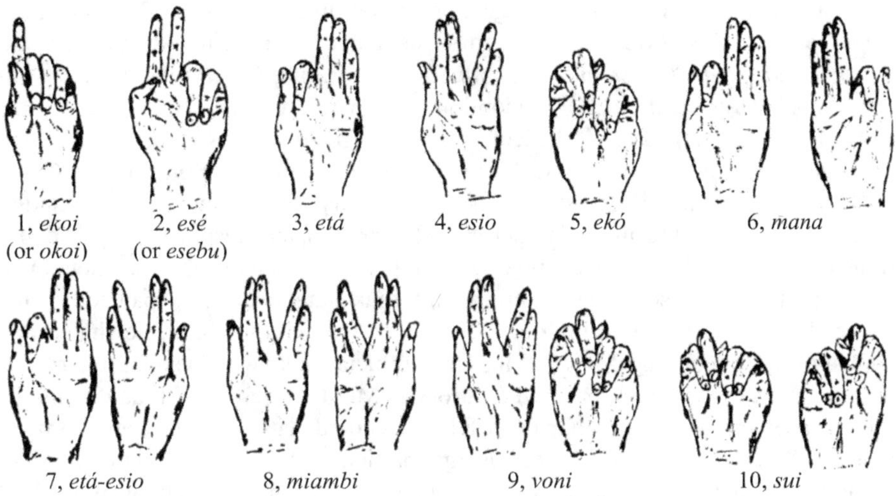

Fig. 22.1 Mongwandi finger-counting. The pairs for *one-two* and *two-four* are displayed on a single hand, while those for *three-six*, *four-eight*, and *five-ten* involve both hands. Adapted from Maes (1911, Fig. 2, unnumbered page) and labeled with vocabulary from Thonner (1910, pp. 87 and 89). Image in the public domain

A finger-counting system similar to the one hypothesized by Kurakichi was attested in early twentieth-century Africa: The Mongwandi, a people of the Mongala River region in what is now the Democratic Republic of the Congo, were described as using a finger-counting system based on pairing (Fig. 22.1).[9] In the Mongwandi case, the pairs *one-two* and *two-four* were formed with one hand, while *three-six*, *four-eight*, and *five-ten* involved both hands. However, the Mongwandi signs for *five-ten* bent the fingers, rather than extending them in the way Kurakichi proposed for Native Japanese, and the associated Mongwandi number-words had no hint of doubling for any of the pairs.[10]

If paired finger-counting was indeed the source of the paired numerals in Native Japanese, no trace of any such system remains in evidence today. If there were such a system, it would most likely have existed a very long time ago, in order for it to have disappeared so completely and for the modern (non-pairing) system to have become so widely prevalent. There's also no method for estimating the age of the number-words from the language itself, as Japanese is not clearly related to any of the major language families spoken in mainland Asia.[11] This means there's no comparative vocabulary to help assess when the number-words might have emerged. And the number-words themselves could always be older. Japanese emerged as a distinct language around 300 BCE, but vocabulary like the number-words could

[9] Flegg (1989); Maes (1911); Thonner (1910).

[10] Thonner (1910).

[11] Frellesvig (2010); Miller (1980); Shibatani (1990).

have originated in whatever language was spoken in Japan during the prehistoric Jōmon era (10,500–300 BCE),[12] named for its distinctive cord-marked pottery.

Japan adopted Chinese numerals, number-words, and calculating methods, but not Chinese finger-counting. The adoption of the written numerals, number-words, and calculating methods isn't difficult to explain, given their usefulness in the kind of economic, military, and political interactions that were going on between China and Japan at the time.[13] Yet despite these adoptions, despite the possible relation between the numeral forms and finger-signs (Chap. 20), and despite the high-fidelity transmission that characterizes finger-counting within societies (as was mentioned in Chap. 18), Chinese finger-counting isn't found anywhere in Japan today. It's interesting that finger-counting doesn't share the same potential to be shared and spread as the calculating devices. This same reluctance slowed the adoption of calculating with written notations in the European Middle Ages,[14] and today it inhibits the replacement of the familiar QWERTY keyboard with more efficient forms.[15] Once we have become proficient at using a particular device that functions well—including the hand—we are often not inclined to learn a new one and its related set of movements for small gains of efficiency.

Ainu numbers are formed by "over-counting."[16] The Ainu people live in Hokkaido, the northernmost large island in the Japanese archipelago, as well as the islands in the Sea of Okhotsk and nearby coastal Russia. Like the Japanese language, Ainu is a linguistic isolate, unrelated to any other known language, living or extinct.[17] As mentioned in Chap. 11, over-counting forms numbers by anticipating a completed larger bundle and adding toward it. Thus, a number like *six* isn't *four from ten*, back-counting that subtracts *four* from *ten*; rather, a number like *six* is *four to* (or *before*) *ten*, which anticipates a completed bundle of *ten* and sees the number *six* as needing *four* more to complete it.[18] Proto-Ainu numbers (not shown) reveal the pattern of formation even more clearly; in the contemporary version (shown), sound changes have made the over-counting pattern more difficult to detect.[19]

Like many of the world's minority languages today, Ainu is at risk of extinction. Its numbers are rarely used, replaced in daily life by number-words from languages like Japanese.[20] Ainu numbers count up to *thousand—ashikne shine wan hot ne*, meaning *five ten-twenties* (5 × (10 × 20))[21]—an amount suggesting significant use and elaboration. However, there's also evidence to suggest these numbers have

[12] Shibatani (1990).

[13] Wang (2023).

[14] Reynolds (1993); Stone (1972).

[15] Chrisomalis (2020).

[16] Menninger (1992).

[17] Shibatani (1990).

[18] Sidwell (1999).

[19] Sidwell (1999).

[20] Refsing (2013).

[21] Batchelor (1889), p. 22.

Table 22.2 Ainu numbers (compiled from Refsing, 2013; Sidwell, 1999, pp. 255–256; also see Batchelor, 1889)

Nr.	Ainu	Meaning
1	*shinep*	*one*
2	*tup*	*two*
3	*rep*	*three*
4	*inep*	*four*
5	*ashiknep*	*five* ("aske" = "hand")
6	*iwanbe*	(*four to ten*)
7	*arawanbe*	(*three to ten*)
8	*topesanbe*	(*two to ten*), or (*lacking in two*)
9	*shinepesanbe*	(*one to ten*), or (*lacking in one*)
10	*wanbe*	("two sides")

never been used with great frequency. Ainu is usually described as a vigesimal system (Table 22.2), and indeed, the numbers up to *two-hundred* (10 × 20) are formed by multiplying *twenty*; however, above *two-hundred*, numbers up to *thousand* are formed as multiples of *two-hundred*, suggesting a decimal organizing principle.[22] Consistent with numbers that aren't used very often, the highest productive Ainu term is *twenty*,[23] and productive terms haven't emerged at the expected places—*four-hundred* (or *twenty squared*) for a vigesimal system; *two-hundred* where vigesimal formation shifts to decimal.

On thc island of Sakhalin, now a Russian territory, a fully decimal version of Ainu numbers was recorded; this system may have reflected an archaic practice of counting sable-skins, since pelts were a common form of tribute paid to the Manchu (Manchurian) occupiers during the two centuries the Ainu were under their rule.[24]

References

Batchelor, J. (1889). *An Ainu-English-Japanese dictionary and grammar*. Hokkaido Cho.

Chrisomalis, S. (2020). *Reckonings: Numerals, cognition, and history*. MIT Press.

Flegg, G. (1989). *Numbers through the ages*. Macmillan.

Frellesvig, B. (2010). *A history of the Japanese language*. Cambridge University Press.

Kurakichi, S. (1937). The Japanese numerals. *Memoirs of the Research Department of the Tōyō Bunko (The Oriental Library), 9*, 1–78.

Laufer, B. (1917). The vigesimal and decimal systems in the Ainu numerals: With some remarks on Ainu phonology. *Journal of the American Oriental Society, 37*, 192–208.

Maes, J.-M. (1911). La numération chez les Mongwandi [Numeration among the Mongwandi]. *La Revue Congolaise, 2*, 68–69.

[22] Batchelor (1889), pp. 20–22.

[23] Batchelor (1889).

[24] Laufer (1917).

Menninger, K. (1992). *Number words and number symbols: A cultural history of numbers* (P. Broneer, Trans.). Dover Publications.

Miller, R. A. (1969). The Altaic numerals and Japanese. *Journal-Newsletter of the Association of Teachers of Japanese, 6*(2), 14–29.

Miller, R. A. (1980). *Origins of the Japanese language; lectures in Japan during the academic year 1977–78*. University of Washington Press.

Nishiyama, Y. (2013). Counting with the fingers. *International Journal of Pure and Applied Mathematics, 85*(5), 859–868.

Refsing, K. (2013). Ainu, Kuril Islands, Hokkaido, Japan. In E. Chan (Ed.), *Numeral systems of the world's languages*. Max Planck Institute. https://lingweb.eva.mpg.de/channumerals/Ainu.htm

Reynolds, B. E. (1993). The algorists vs. the abacists: An ancient controversy on the use of calculators. *College Mathematics Journal, 24*, 218–223.

Shibatani, M. (1990). *The languages of Japan*. Cambridge University Press.

Sidwell, P. (1999). The Austroasiatic numerals "1" to "10" from a historical and typological perspective. In J. Gvozdanović (Ed.), *Numeral types and changes worldwide* (pp. 253–271). Mouton de Gruyter.

Stone, W. E. (1972). Abacists versus algorists. *Journal of Accounting Research, 10*(2), 345–350.

Thonner, F. (1910). *Vom Kongo zum Ubangi: Meine zweite Reise in Mittelafrika [From Congo to Ubangi: My second trip to Central Africa]*. Dietrich Reimer.

Wang, X. (2023). The establishment of the East Asian international order in the seventh century. In P. Bu & K. Shinichi (Eds.), *The history of China-Japan relations: From ancient world to modern international order* (pp. 57–88). Springer Nature.

Chapter 23
Korean Numbers and Numeral Types

Abstract Korea borrowed the Chinese system of writing, including its numerals and calculating methods, in the third or fourth century CE. During the reign of King Sejong the Great, a unique and rare "featural" type of alphabet was developed for writing the Korean language, invented either by the king himself or by scholars he commissioned. This alphabet, known as *Han'gŭl*, bases the shapes of its letters on "articulatory features," the movements made by the mouth in pronouncing Korean words. As a featural alphabet, Han'gŭl lacks *semasiographic* signs (these are signs that lack pronunciation clues), including numerals. The Han'gŭl numerals are in fact the Korean words for numbers, written glottographically (that is, written lexically or phonographically in a way that provides clues for pronouncing the Korean number-words). A parallel in English would write numbers as words: "one," "two," "three." The Han'gŭl numerals are unique, since numeral systems around the world are semasiographic rather than lexical, phonetic, or glottographic. In lacking semasiographic numerals, the Korean writing system joins the small minority of writing systems that lack such signs.

Keywords Han'gŭl · Featural alphabet · Semasiography · Conventions · Glottographic writing

Sandwiched as it is between the Japanese archipelago and Chinese mainland, Korea has a long history of interacting—both favorably and not—with its neighbors, particularly China. About a thousand years after the time of the oracle bones and long after the Shang Dynasty had been succeeded by the Zhou and Qin, the Han Dynasty (202 BCE–220 CE, as briefly interrupted by the Xin Dynasty, 9–23 CE) invaded and then colonized Korea in 109 BCE. In the same way Japan would later adopt and adapt the Han system of writing and Chinese vocabulary in the eighth century, Korea adopted them in the third or fourth century, adapting "*hanja*" to the Korean language and developing a Chinese-influenced "Sino-Korean" language.[1]

[1] Taylor and Taylor (2014).

K. A. Overmann, *Cultural Number Systems*, Interdisciplinary Contributions to Archaeology, https://doi.org/10.1007/978-3-031-83383-0_23

The invention of the Korean alphabet was a scientific triumph. Writing Korean with Chinese characters was so difficult that literacy was effectively limited to the educated (male) nobility. In 1446 CE, King Sejong the Great, the fourth monarch of the Chosŏn Dynasty, wanted to improve literacy among the common people; accordingly, he either himself invented or commissioned his scholars—an impressive think tank known collectively as the Hall of Worthies—to create Han'gŭl, an alphabet tailored to the Korean language.[2] The shapes of the letters for consonants were based on the movements made by the mouth in pronouncing them: not just the lips, but also the tongue, mouth cavity, and vocal cords; the result was a "featural" writing system, a rare type of alphabet in which the letters represent articulatory features, rather than sounds.[3] Han'gŭl letters were also grouped as syllables, rather than being arranged along straight lines; this gained them the benefit of being recognized as syllables[4] and created the illusion they were more similar to Chinese than they actually were.

As a featural system for representing spoken language, Han'gŭl lacks any numerals. To understand this, we first need to explore the different ways of representing numbers. On the top row of Fig. 23.1, "semasiographic" forms represent numbers through the quantity of their elements. Semasiography—a name that means "writing with signs"—communicates information without involving language.[5] Numerals like the Western symbols 0–9 are a classic example of a semasiographic system (though the numerals 4 and higher are conventions, described in the next paragraph), and musical notations are another. Semasiographic forms represent a number like 3 by being composed of three elements, and they do so irrespective of whether they are material objects (e.g., fingers, tallies, knots, beads, and tokens) or written notations (e.g., cuneiform, Roman numerals, and the small numbers 1 through 3 in Chinese and Western numerals).[6] Because sound values are not involved, these forms can be read in any language.[7] This very quality has enabled the use of Western numerals around the world—across hundreds if not thousands of languages and writing systems. It's also the reason why we can use Roman numerals today without needing to learn their names in Latin.

In the middle row of Fig. 23.1, semasiographic forms represent numbers through conventions, rather than their quantity of elements. That is, these forms mean a number like 10 or 4 through a social agreement that established these meanings for these forms, irrespective of whether they are material objects (e.g., tokens) or written notations (e.g., Linear A, Chinese, Roman, Dogri, Khmer, Chinese, and Western numerals). Very often, such conventions are created to represent a bundled quantity, a single sign that consolidates and represents an agreed-upon amount of

[2] Foundation (2010); Lee (1997).

[3] Foundation (2010); Ramsey (1997).

[4] Reischauer and Fairbank (1960).

[5] Powell (2009).

[6] Overmann (2022).

[7] Chrisomalis (2010).

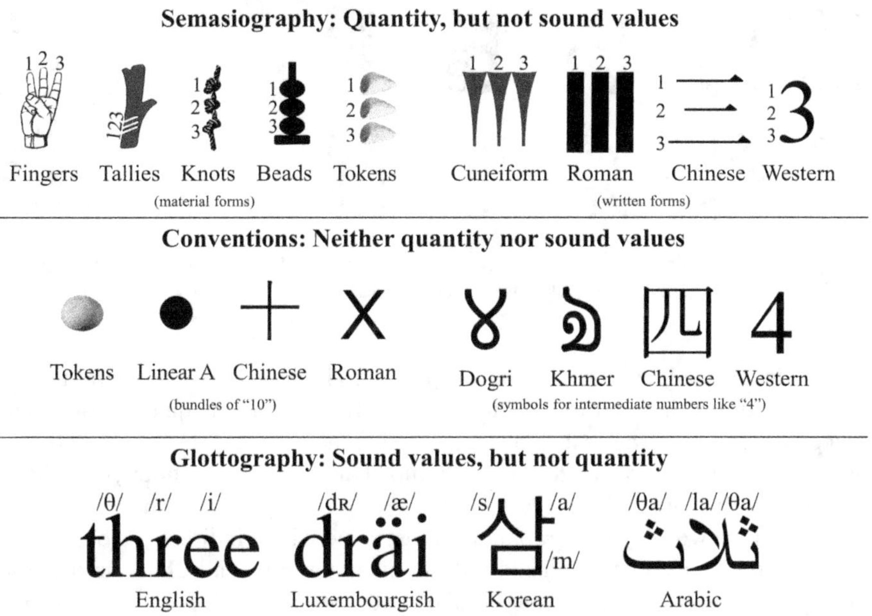

Fig. 23.1 Three ways to represent numbers: semasiography, conventions, and glottography. Image by the author

lower-quantity units (10 is typical for decimal systems). Conventions can emerge over time as a function of how writing systems change, as was the case for the familiar Western numerals,[8] or they can be created by a particular society, as was likely the case for the Chinese numerals. In either case, these signs emerge from numerical systems involving unwritten precursors like fingers and tallies. As these forms involve neither quantity nor sound values, their meanings are simply learned, though they can also be figured out—assuming there is enough of a sample to permit analysis—because they embody numerical relations. For example, in the Minoan Linear A system, a solid circle meant 10,[9] so there shouldn't be more than nine of the unit signs, as ten units would be replaced by the sign for 10.

Finally, on the bottom row of Fig. 23.1, "glottographic" forms represent words for numbers by visually depicting their sound values (or, in the case of Han'gŭl, the articulatory features that produce specific sounds). Glottographic signs visually depict the sounds of speech in some way, either as morphemes (a unit of speech roughly analogous to words), syllables, individual consonant and vowel sounds (also known as phonemes), or articulatory features. Since these signs mean sounds, glottographic writing is inextricably tied to whatever particular language it represents (as shown in the figure, English, Luxembourgish, Korean, and Arabic), and

[8] Overmann (2021).

[9] Packard (1974).

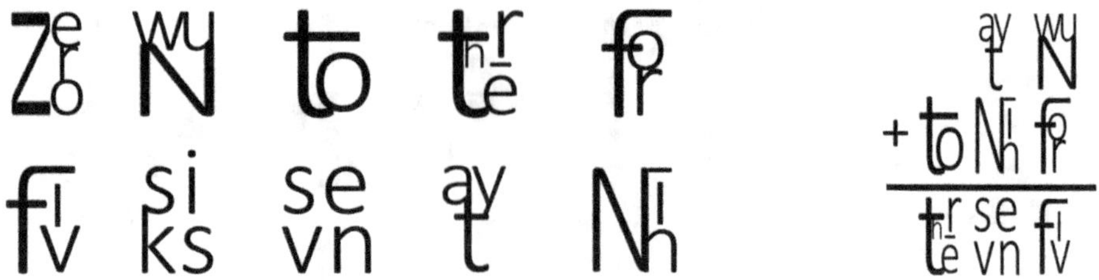

Fig. 23.2 The English numbers *zero* through *nine*, (left) spelled phonetically and presented as Han'gŭl style "numerals" and (right) used to add 81 + 294 = 375. Image by the author

knowledge of the language, as well as specific training and practice with the script, are both required in order to be able to read it. In addition, the quantity of elements in a glottographic numerical sign is unrelated to the number-word expressed. To get an idea of how truly different glottographic "numerals" are, similar (hypothetical) ones for English are shown in Fig. 23.2.

Around the world, glottographic forms descend, albeit indirectly, from the original writing systems that emerged in Mesopotamia, Egypt, and China. In the case of Han'gŭl, the alphabet was created in the context of an existing writing system, the *hanja* borrowed from the Han; this means that at the time Han'gŭl was invented, both the idea of writing and its extensive practice had been in place for centuries. And because the Han writing system had to be adapted to express the Korean language, the context for inventing Han'gŭl also included the specific investigation of the relation between written signs and the sounds of speech, conducted by examining phonetic scripts written in the Chinese, Manchurian, Mongolian, Japanese, and possibly Indian languages.[10] Because the result reflects an astonishing amount of analysis and insight, Han'gŭl has aptly been praised as "the most scientific writing in general use in any country."[11] Despite its advantages, the script was not initially a success, as the nobility resisted it as too common and anti-Chinese; it was also banned twice, first in 1504 by Yŏnsan'gun, a tyrant so brutal he was deposed, exiled, and permanently downgraded to prince, and again in 1938 by Japan (which ruled Korea 1910–1945) to suppress Korean culture.[12]

Han'gŭl "numerals" are Korean words for numbers, written glottographically. The number-words themselves are either Sino-Korean or Native Korean (Table 23.1). The Sino-Korean numerals and number-words, which are simpler and shorter, are used for "dates, money, addresses, phone numbers," the numbers *one hundred* and higher, and calculating, while the longer, more complex Native Korean numbers are used for counting to *ninety-nine* and for expressing human age.[13] The Native Korean words for higher numbers have been replaced, either by Sino-Korean number-words

[10] Taylor and Taylor (2014).

[11] Reischauer and Fairbank (1960), p. 435.

[12] Taylor and Taylor (2014).

[13] Fuson and Kwon (1992a, 1992b); Lee and Kang (2024).

Table 23.1 Korean number-words and "numerals" (adapted from Lee & Kang, 2024)

English words	Chinese numerals	Sino-Korean		Native Korean	
		Words	*Han'gŭl*	Words	*Han'gŭl*
one	一	*il*	일	*hana*	하나
two	二	*i*	이	*dul*	둘
three	三	*sam*	삼	*set*	셋
four	四	*sa*	사	*net*	넷
five	五	*o*	오	*daseot*	다섯
six	六	*yuk*	육	*yeoseot*	여섯
seven	七	*chil*	칠	*ilgop*	일곱
eight	八	*pal*	팔	*yeodeol*	여덟
nine	九	*gu*	구	*ahop*	아홉
ten	十	*sip*	십	*yeol*	열

Note: The numbers in all the columns are glottographic, with the exception of the Chinese column, where the first three numerals are semasiographic signs that express number through their quantity of elements and the remaining numerals are semasiographic conventions

(as for example, 121 is *payk-i-sip-il*, [*one*] *hundred two tens one* in Sino-Korean) or by compounds that combine them (as for example, 121 is *payk-sumul-hana*, [*one*] *hundred twenty-one*, where [*one*] *hundred* is Sino-Korean and *twenty-one* is Native Korean).[14]

In lacking numerical notations and expressing numbers as number-words written glottographically, Han'gŭl joins a handful of other scripts: "the Ogham script of Ireland, the Canaanite script, the early alphabets of Asia Minor such as Carian and Phrygian, and the indigenous scripts of the Philippines."[15] The emphasis on the articulatory features involved in producing the sounds of speech was likely the reason why semasiographic forms (that is, forms without any clues to sound values, including the articulatory features that produce them) were not developed for Han'gŭl.

The type of numeral may affect numeracy, but we don't know whether it does or how. Visually complex characters like Chinese and Han'gŭl are more costly in terms of brain processing,[16] and the Han'gŭl characters are not just complex, they are also glottographic rather than semasiographic. Yet the Sino-Korean number-words used to teach numbers to Korean children are "transparent" because the words for higher numbers are formed from smaller ones without changes in their form or pronunciation; this regularity makes numerical structure more apparent, which makes learning and using numbers easier.[17] Beyond written symbols and spoken number-words

[14] Sohn (1999).

[15] Chrisomalis (2010), p. 21.

[16] Cantlon and Brannon (2007); Tan et al. (2005).

[17] Fuson and Kwon (1992a, 1992b).

are differences in the algorithms used and how highly education is valued; these matters make it challenging to compare numeracy across cultures.[18]

References

Cantlon, J. F., & Brannon, E. M. (2007). Adding up the effects of cultural experience on the brain. *Trends in Cognitive Sciences, 11*(1), 1–4.

Chrisomalis, S. (2010). *Numerical notation: A comparative history*. Cambridge University Press.

Foundation, K. (2010). *Hangeul: Korea's unique alphabet*. Seoul Selection.

Fuson, K. C., & Kwon, Y. (1992a). Korean children's understanding of multidigit addition and subtraction. *Child Development, 63*(2), 491–506.

Fuson, K. C., & Kwon, Y. (1992b). Learning addition and subtraction: Effects of number words and other cultural tools. In J. Bideaud, C. Meljac, & J.-P. Fischer (Eds.), *Pathways to number: Children's developing numerical abilities* (pp. 283–306). Lawrence Erlbaum Associates.

Lee, K.-M. (1997). The inventor of the Korean alphabet. In K.-R. Young-Key (Ed.), *The Korean alphabet: Its history and structure* (pp. 11–30). University of Hawai'i Press.

Lee, Y., & Kang, J. J. (2024). Numbers in Korean. *Omniglot: The Online Encyclopedia of Writing Systems and Languages*. https://www.omniglot.com/language/numbers/korean.htm

Overmann, K. A. (2021). Writing system transmission and change: A neurofunctional perspective. In G. Gabriel, K. A. Overmann, & A. Payne (Eds.), *Signs – sounds – semantics. Nature and transformation of writing systems in the ancient Near East* (pp. 93–116). Ugarit-Verlag.

Overmann, K. A. (2022). Early writing: A cognitive archaeological perspective on literacy and numeracy. *Visible Language, 56*(1), 8–44.

Packard, D. W. (1974). *Minoan Linear A*. University of California Press.

Powell, B. B. (2009). *Writing: Theory and history of the technology of civilization*. Wiley-Blackwell.

Ramsey, S. R. (1997). The invention of the alphabet and the history of the Korean language. In K.-R. Young-Key (Ed.), *The Korean alphabet: Its history and structure* (pp. 131–143). University of Hawai'i Press.

Reischauer, E. O., & Fairbank, J. K. (1960). *East Asia: The great tradition*. Houghton Mifflin.

Sohn, H.-M. (1999). *The Korean language*. Cambridge University Press.

Tan, L.-H., Laird, A. R., Li, K., & Fox, P. T. (2005). Neuroanatomical correlates of phonological processing of Chinese characters and alphabetic words: A meta-analysis. *Human Brain Mapping, 25*(1), 83–91.

Taylor, I., & Taylor, M. M. (2014). *Writing and literacy in Chinese, Korean and Japanese* (2nd ed.). John Benjamins.

[18] Fuson and Kwon (1992a, 1992b).

Chapter 24
Numbers in the Philippines

Abstract Early Spanish missionaries recorded the number-words of Old Tagalog before these had been significantly influenced by the Spanish language. Old Tagalog numbers were constructed with an unusually complex form of over-counting, with four main rules for naming ten sets of *ten* for each power of *ten*. The Filipinos were literate and numerate long before the Spanish arrived in 1521. Centuries earlier, the Filipinos had borrowed scripts from India that contained numerals, and an Indic script called *baybayin* was used to write Old Tagalog. However, the Filipinos did not use the Indic numerals, instead writing out the Old Tagalog numbers in baybayin as pronounceable words. Most systems of numerical notations accumulate, forming new numbers by adding to a lower bundle (as VI in Roman numerals adds I to V). A few numeral systems back-count, forming new numbers by subtracting from a higher bundle (as IV in Roman numerals subtracts I from V). No known system of numerical notations represents over-counting. Ignoring the available Indic numerals suggests that Old Tagalog numbers were written out phonetically because this method was better at representing their structural logic.

Keywords Austronesian language · Tagalog · Baybayin · Over-counting · Phonetically written numerals

The Philippines are an archipelago of more than 7000 islands in the western Pacific, as bounded by the Celebes, Philippine, and South China Seas. Though geographically located in Oceania, it's often assigned to Asia—a part of Maritime Southeast Asia, along with Indonesia, Malaysia, Singapore, Brunei, and East Timor—because of historical and cultural ties.[1] After the Portuguese explorer Ferdinand Magellan made contact with the islands in 1521, the archipelago was (re-)named in honor of Philip II, then king of Spain and Portugal, in 1543. Here the focus is on Tagalog, brought to the Philippines by Austronesian settlers about 3500 years ago and today

[1] Hall (2011).

K. A. Overmann, *Cultural Number Systems*, Interdisciplinary Contributions to Archaeology, https://doi.org/10.1007/978-3-031-83383-0_24

Table 24.1 Number-words in five Austronesian languages (Bouwer, 2009; Cook, 1777, Vol. II, p. 364; de San Ioseph, 1610, p. 264; Lee, 1820, p. 17; Li, 1993)

Nr.	Pazeh (Taiwan)	Malagasy (Madagascar)	Old Tagalog (Philippines)	Rapanui (Rapa Nui)	Māori (Aotearoa)
1	*ida*	*ray*	*ysa*	katta'haee	*ko táhi*
2	*dusa*	*roa*	*dalava*	'rooa	*ka dúa*
3	*turu*	*telo*	*tatlo*	'toroo	*ka tódu*
4	*supat*	*efatra*	*apat*	'häa, fäa	*ka wá*
5	*xasəp*	*dimy*	*lima*	'reema	*ka díma*
6	*xasəbuza*	*enina*	*anim*	'honoo	*ka óno*
7	*xasəbidusa* (5 + 2)	*fito*	*pito*	'heedoo	*ka wítu*
8	*xasəbituru* (5 + 3)	*valo*	*valo*	varoo	*ka wádu*
9	*xasəbisupat* (5 + 4)	*sivy*	*siyam*	heeva	*ka íwa*
10	*isit*	*folo*	*poló*	atta'hooroo	*ka nga údu*

the national language for a country that encompasses around 150 languages.[2] We are interested in the number-words of Old Tagalog, as documented by early Spanish missionaries and Tomás Pinpin, an Indigenous printer of early books in Tagalog.[3] Old Tagalog number-words differ from today's Tagalog number-words, which have been significantly altered through Spanish contact.

Tagalog number-words have Austronesian roots. Austronesian is a family of languages spoken in an area delimited by Taiwan to the north, Madagascar to the west, and Polynesia to the east (Rapa Nui, also called Easter Island) and south (Aotearoa, also called New Zealand). The Philippines occupy a middle position historically critical to regional trade. Austronesian number-words are shown in Table 24.1, as recorded in early sources (Old Tagalog, Rapanui, Māori) and by contemporary researchers (Pazeh and Malagasy). Except for the formations *five* and higher in Pazeh and despite the mix of spellings and diacritics, the words are recognizable as shared. The word *lima*, meaning *hand*, is used for the number *five* throughout the Pacific, but interestingly, *lima* does not mean *hand* in Tagalog, where the word for *hand* is instead *kamáy*.[4]

Old Tagalog had an unusual way of building its numbers. Austronesian languages have decimal numbers, and the majority—including Tagalog today—build their higher numbers similarly to Spanish and English: *veinte* (*twenty*) is *two tens*, *veintiuno* (*twenty-one*) is *two tens [and] one*, and so on, with productive terms at *ten*, *hundred*, and *thousand*. In Old Tagalog, the decades, hundreds, thousands, and

[2] Schachter and Reid (2018).

[3] Woods (1992).

[4] Potet (1992).

Table 24.2 Powers of *ten* in Old Tagalog (data for 10^1 to 10^5 from de San Ioseph, 1610; data for 10^6 to 10^8 from De Noceda and de San Lucar, 1754)

Nr.	Power	Old Tagalog	Nr.	Power	Old Tagalog
10	10^1	*sangpouo*	100,000	10^5	*sangyota*
100	10^2	*sangdaan*	1000,000	10^6	*angao angao* or *gatos*
1000	10^3	*sanglibo*	10,000,000	10^7	*cati*
10,000	10^4	*sanglacsa*	100,000,000	10^8	*bahala*

so on were formed similarly (Rule 1): *twenty* was *dalavangpouo* or *two tens*, *one hundred* was *sangdaan*, and *one thousand* was *sanglibo* (Table 24.2).[5]

Within the sets, however, Old Tagalog numbers were formed quite differently. Numbers within the first set were "more than" their *decade*, *hundred*, or *thousand* (Rule 2), so *eleven* was *labin isa*, "one [unit] in excess [of ten]"; 110 was *labisadaan ysa*, "one [ten] in excess of one hundred"; and 1100 was *labisalibon ysa*, "one [hundred] in excess of one thousand."[6] *One hundred and ten* could also be said *labisadaan sangpouo*, and to say *one hundred and one*, the term *labisadaan sangsalop* was used.[7] Numbers within sets two through eight anticipated the next higher set (Rule 3, the use of over-counting), so *twenty-one* was *maycatlon ysa*, "one [unit] in the third set of ten"; 210 was *maycatlong daan ysa*, "one [ten] in the third set of hundred"; and 2100 was *maycatlong libon ysa*, "one [hundred] in the third set of thousand."[8] Numbers within the tenth (or last) sets completing a power anticipated the next higher one (Rule 4, also over-counting): *hundred* for decades, *thousand* for hundreds, *ten thousand* for thousands.[9]

Understand how numbers were constructed in Old Tagalog is easier when they are visualized as ten sets for each power of *ten* (Fig. 24.1). This was perhaps the kind of layout once used with heaps of pebbles ("contaban con piedrecìllas, haciédolas montoncìllos"[10]) and counting sticks ("palillos de contar"[11]).

The way Old Tagalog numbers were constructed resembles the over-counting used in Papua New Guinea, suggesting the possibility of borrowing. However, Kakoli numbers are constructed on *four* and *thirty-two*, rather than *ten*; Kakoli is a language in the Trans-New Guinea family, not Austronesian;[12] and how much the highland New Guineans might have come into contact with the seagoing Tagalogs is unknown. Rule 4 also resembles the way nines are constructed Hindi (Chap. 25), again suggesting possible regional influences.

[5] De San Ioseph (1610).

[6] De San Ioseph (1610).

[7] De San Ioseph (1610).

[8] De San Ioseph (1610).

[9] Potet (2022).

[10] De San Antonio (1738), Part I, Book I, Chapter XLV, p. 167.

[11] De Noceda and de San Lucar (1754), pp. 75, 123, 269, 361.

[12] Bowers and Lepi (1975); Head (2008).

	Ten thousands	Rule 1	Thousands	Rule 1	Hundreds	Rule 1	Tens	Rule 1	Ten sets
Rule 4	**"in the last set for 100,000" (90,000s)**	1×100,000	"in the last set for 10,000" (9000s)	1×10,000	"in the last set for 1000" (900s)	1×1,000	"in the last set for 100" (90s)	1×100	10
Rule 3	"in the 9th set of 10,000" (80,000s)	9×10,000	"in the 9th set of 1000" (8000s)	9×1,000	"in the 9th set of 100" (800s)	9×100	**"in the 9th set of 10" (80s)**	9×10	9
	"in the 8th set of 10,000" (70,000s)	8×10,000	"in the 8th set of 1000" (7000s)	8×1,000	"in the 8th set of 100" (700s)	8×100	"in the 8th set of 10" (70s)	8×10	8
	"in the 7th set of 10,000" (60,000s)	7×10,000	**"in the 7th set of 1000" (6000s)**	7×1,000	"in the 7th set of 100" (600s)	7×100	"in the 7th set of 10" (60s)	7×10	7
	"in the 6th set of 10,000" (50,000s)	6×10,000	"in the 6th set of 1000" (5000s)	6×1,000	"in the 6th set of 100" (500s)	6×100	"in the 6th set of 10" (50s)	6×10	6
	"in the 5th set of 10,000" (40,000s)	5×10,000	"in the 5th set of 1000" (4000s)	5×1,000	"in the 5th set of 100" (400s)	5×100	"in the 5th set of 10" (40s)	5×10	5
	"in the 4th set of 10,000" (30,000s)	4×10,000	"in the 4th set of 1000" (3000s)	4×1,000	"in the 4th set of 100" (300s)	4×100	"in the 4th set of 10" (30s)	4×10	4
	"in the 3rd set of 10,000" (20,000s)	3×10,000	"in the 3rd set of 1000" (2000s)	3×1,000	"in the 3rd set of 100" (200s)	3×100	"in the 3rd set of 10" (20s)	3×10	3
	"in the 2nd set of 10,000" (10,000s)	2×10,000	"in the 2nd set of 1000" (1000s)	2×1,000	"in the 2nd set of 100" (100s)	2×100	"in excess of 10" [Rule 2]	2×10	2
Rule 2	"in excess of 10,000"	1×10,000	"in excess of 1000"	1×1,000	**"in excess of 100"**	1×100	(units) [rules not applicable]	1×10	1

Fig. 24.1 Old Tagalog numbers organized as sets of *ten*. Bold text in highlighted cells shows the formation of 96,185, *maikayútańg maikapitóng líboñg labí sa raáñg maikasiyám na líma* ("in the last set [of 10,000] for 100,000" = 90,000, Rule 4) plus ("in the seventh set of 1,000" = 6000, Rule 3) plus ("in excess of 100" = 100, Rule 2) plus (5 "in the ninth set [of 10]" = 85, Rule 3). Image by the author based on data from Cordial (2023) and Potet (2022)

Within decades of the Spanish arrival, Spanish words for numbers began to influence the numbers in Tagalog. Today, Tagalog numbers are constructed like Spanish numbers: for example, *veintiuno* and *dalawam'puʔt ʔi'sa* are both "two tens and one."[13]

Old Tagalog counted to very high numbers. The oldest dictionaries give terms for numbers up to the fifth power of *ten*, later ones the eighth power of *ten* (Table 24.2). Very large numbers and complex geometric concepts are consistent with Tagalog accomplishments and skills in domains like timekeeping, trade, and shipbuilding.[14] Early Spanish observers, however, either ignored or severely underestimated Tagalog numeracy, and they were apparently mystified about why the Tagalogs needed such high numbers in the first place. In 1703, for example, the Spanish missionary Gaspar de San Agustin wrote that although the Tagalogs could count to *one hundred thousand* and infinity, they were ill suited for arithmetic ("los Tagalos son poco aritméticos"), and they were unreliable in counting and bad at arithmetic ("Pero los tagalos en el contar son varios y malos aritméticos").[15] A modern researcher explained the Spanish mindset as follows: "One cannot imagine that such high powers of ten were of any use" to the Tagalogs in their day-to-day pursuits; such large numbers were perhaps "necessary only in speculative calculations involving very large numbers with no practical applications."[16]

The highest numbers in Table 24.2 (*ten* to the sixth power and higher) appear as shown in a 1754 dictionary compiled by two Jesuit priests, Juan de Noceda and Pedro de San Lucar.[17] *Gatos* may have been given the meaning *billion* (*ten* to the

[13] Rubino (2006).

[14] Manapat (2011).

[15] De San Agustin (1879), pp. 115, 117; also see Manapat (2011); Morrow (2010).

[16] Potet (1992), pp. 178–179.

[17] De Noceda and de San Lucar (1754); also see discussion in Morrow (2010).

ninth power) in the 1900s.[18] The idea that numbers higher than *ten* to the fifth power were later additions is consistent with the unsettled nature of their values, as well as an observation made in 1610 by the Spanish missionary Francisco de San Ioseph: *bahala* meant, "What do I know? I leave it up to you [Bahala ka]. What can I do? Of these things one can no longer conceive."[19] On the other hand, it's possible de San Ioseph confused *bahálà*, meaning *hundred million*, with a homonym meaning "responsibility."[20]

Filipinos were numerate and literate long before the Spanish arrived. However, their system of writing lacked numerals, semasiographic symbols like the Roman and Western numerals used by the Spanish. As noted by the missionary Juan Francisco de San Antonio in 1738, "there are no arithmetical numbers among their characters, such as we use" ("en sus Caractères no ay Números arifthméticos, como nosotros usamos").[21] Numbers in Old Tagalog were instead written phonetically. For example, in the 1593 *Doctrina Christiana*, one of the earliest publications in Tagalog, *fourteen* is *labin apat* (in Tagalog) and ᜎᜊᜒ ᜀᜉ (*la-bi a-pa* in baybayin, the Tagalog script).[22] As we saw with Korean Han'gŭl (Chap. 23), such phonetically written numerals are rare among the world's number systems.

Baybayin is an *abugida*, a type of syllabary that emphasizes consonants and reduces vowels to marks indicating pronunciations. While the origins of baybayin are unclear and disputed, it is generally classified as a Brahmic or Indic script because of its similarities with scripts of that type found in India, Indonesia, and Malaysia. Along with the Buddhist and Hindu religions, Indic scripts spread through sea routes connecting Maritime Southeast Asia in the last centuries BCE or first centuries CE.[23] This origin is interesting because Indic scripts do contain numerals,[24] while baybayin doesn't. Moreover, because of their strategic location for regional trade, the Tagalogs came into contact with many numeral systems.[25] This would have given them a range of straightforwardly decimal options, as Indic numerals can be additive like Roman numerals or multiplicative like Chinese numerals.[26]

However, there's a likely reason why the Tagalogs adopted none of the numeral options available to them. As was noted in an earlier section, no known numeral system is able to capture the complexity of over-counting, even when it is used far more sparingly than was true of Old Tagalog. Given the complex manner in which Old Tagalog numbers were formed (Fig. 24.1), writing numbers phonetically might have been easier—more revealing of their structural logic—compared to any of the available notational forms, regardless of their organization.

---

[18] Morrow (2010).

[19] De San Ioseph (1610), p. 266; as translated in Manapat (2011).

[20] Potet (2022).

[21] De San Antonio (1738), Part I, Book I, Chapter XLV, p. 167; as translated by Blair and Robertson (1906), p. 364.

[22] De Plasencia (1593).

[23] Gardner (1943).

[24] Chrisomalis (2010).

[25] Potet (2014, 2022).

[26] Chrisomalis (2010).

References

Blair, E. H., & Robertson, J. A. (Eds.). (1906). *The Philippine islands 1493–1898. Vol. XL: 1690–1691* (E. H. Blair & J. A. Robertson, Trans.). Arthur H. Clark Company.

Bouwer, L. (2009). Malagasy (Merina), Madagascar. In E. Chan (Ed.), *Numeral systems of the world's languages*. Max Planck Institute. https://lingweb.eva.mpg.de/channumerals/Malagasy-Merina.htm

Bowers, N., & Lepi, P. (1975). Kaugel Valley systems of reckoning. *Journal of the Polynesian Society, 84*(3), 309–324.

Chrisomalis, S. (2010). *Numerical notation: A comparative history*. Cambridge University Press.

Cook, J. (1777). *A voyage towards the South Pole and round the world, performed in His Majesty's ships, the* Resolution *and* Adventure, *in the years 1772, 1773, 1774 and 1775. Vol. 2, Book III: From Ulietea to New Zealand.* W. Strahan and T. Cadell.

Cordial, J. (2023). Complexity of Old Tagalog counting system. https://medium.com/@jeremiahcordial/would-you-rather-count-in-old-tagalog-ee67d4f3589f

De Noceda, J., & de San Lucar, P. (1754). *Vocabulario de la lengua Tagala, trabaxado por varios sugetos doctos, y graves, y ultimamente añadido, corregido, y coordinado [Vocabulary of the Tagalog language, worked on by several learned and serious subjects, and recently added, corrected, and coordinated].* Nicolas de la Cruz Bagay.

De Plasencia, J. (1593). *Doctrina Christiana, en lengua española ytagala, cor regida por los Religiosos de las ordenes Impressa con licencia, en S. Gabriel de la orden de S. Domigo [Christian Doctrine, in Spanish Tagalog, corrected by the Religious of the Orders Printed with license, in S. Gabriel of the Order of St. Dominic].*

De San Agustin, G. (1879). *Compendio del arte de la lengua Tagala [Compendium of the art of the Tagalog language]* (3rd ed.). Amigos del Pais.

De San Antonio, J. F. (1738). *Chronicas de la apostolica provincia de S. Gregorio de religiosos descalzos de N.S.P.S. Francisco en las islas Philipinas, China, Japon, &c. Parte primera, en qve se inclvye la descripcion de estas islas, qve consagra a la S.C.R. magestad de D. Phelipe V, el animoso, nuestro cathólico rey, y augusto emperador de las Españas, y de las Indias, la misma Santa Provincia, y en su nombre su ministro provincial [Chronicles of the Apostolic Province of St. Gregory of Discalced Religious of Our Lord Francis in the Philippine Islands, China, Japan, &c. First part, in which is included the description of these islands, which consecrates to the Most Holy Majesty of D. Philip V, the courageous, our Catholic King, and August Emperor of Spain and the Indies, the same Holy Province, and in his name his provincial minister].* Juan del Sotillo.

De San Ioseph, F. (1610). *Arte y reglas de la lengua Tagala [Art and rules of the Tagalog language].* Thomas Pinpin.

Gardner, F. (1943). *Philippine Indic studies*. The Witte Memorial Museum.

Hall, K. R. (2011). *A history of early Southeast Asia: Maritime trade and societal development, 100–1500*. Rowman & Littlefield.

Head, J. (2008). Umbu-Ungu, Papua New Guinea. In E. Chan (Ed.), *Numeral systems of the world's languages*. Max Planck Institute. https://lingweb.eva.mpg.de/channumerals/Umbu-Ungu.htm

Lee, S. (1820). *A grammar and vocabulary of the language of New Zealand.* Church Missionary Society.

Li, P. J. (1993). Pazeh, Miaoli county, Taiwan. In E. Chan (Ed.), *Numeral systems of the world's languages*. Max Planck Institute. https://lingweb.eva.mpg.de/channumerals/Pazeh.htm

Manapat, R. (2011). Mathematical ideas in early Philippine society. *Philippine Studies, 59*(3), 291–336.

Morrow, P. (2010). Counting the old way. *Pilipino Express, 6*(10), 1–3.

Potet, J.-P. G. (1992). Numeral expressions in Tagalog. *Archipel, 44*(1), 167–181.

Potet, J.-P. G. (2014). *Baybayin: The syllabic alphabet of the Tagalogs*. Lulu Press.

Potet, J.-P. G. (2022). *Numbers and units in Old Tagalog*. Lulu Press.

Rubino, C. (2006). Tagalog, Philippines. In E. Chan (Ed.), *Numeral systems of the world's languages*. Max Planck Institute. https://lingweb.eva.mpg.de/channumerals/Tagalog.htm

Schachter, P., & Reid, L. A. (2018). Tagalog. In B. Comrie (Ed.), *The world's major languages* (3rd ed., pp. 852–876). Routledge.

Woods, D. L. (1992). Tomás Pinpin and the literate Indio: Tagalog writing in the early Spanish Philippines. *UCLA Historical Journal, 12*, 177–220.

Chapter 25
Numbers in India

Abstract India was home to one of the world's earliest civilizations, the Indus River Valley civilization that flourished between 3300 and 1300 BCE. The system of writing developed there remains undeciphered. Today, India encompasses dozens of writing systems and hundreds of languages and dialects. One of India's 20 official languages is Hindi, a member of the Indo-European language family. In Hindi, the words for numbers up to *ninety-nine* are irregular, a trait thought to impose a significant demand on human memory. Typically, irregularity is confined to the lower numbers, with the smallest, *one* to *about three or four*, always being the most irregular as a consequence of the way the first words for numbers emerge in any language. In its unusual irregularity, Hindi is contrasted with Mandarin Chinese, a language whose numbers are profoundly regular. Regularity provides a transparency thought to aid the learning and acquisition of words for numbers. Some of the region's languages count by twelves rather than tens, creating duodecimal systems of numbers. Examples include Sora and Chepang, languages that are unrelated to one another. Sora numbers additionally involve a vigesimal base.

Keywords Hindi · Transparency · Duodecimal · Sora · Chepang

The subcontinent known today as India is bounded to the south by the Arabian and Laccadive Seas, the Indian Ocean, and the Bay of Bengal. To the north, the subcontinent is separated from the Tibetan Plateau by the world's highest mountain range, the Himalayas, where it neighbors the modern countries of Pakistan, China, Nepal, Bhutan, Bangladesh, and Myanmar. India is not just the world's most populous country and its largest democracy, it is also one of the most geographically and climatically diverse on the planet, spanning tropical forests to arid deserts and alpine tundra and glaciers.[1] It was home to one of the world's earliest civilizations, which flourished in the Indus River Valley between 3300 and 1300 BCE;[2] the system of

[1] Calkins et al. (2024).

[2] Wright (2010).

K. A. Overmann, *Cultural Number Systems*, Interdisciplinary Contributions to Archaeology, https://doi.org/10.1007/978-3-031-83383-0_25

writing developed there, the Harappan script, has yet to be deciphered, and after its decline, "the graphic record of India is virtually a total blank for well over a thousand years."[3] Today, the country encompasses hundreds of languages and dialects, along with dozens of writing systems. India has more than 20 official languages, one of which is Hindi, a member of the Indo-European family related (though distantly) to English.

The Hindi words for numbers are unusually irregular.[4] Irregularity, in and of itself, isn't noteworthy, since there are irregular number-words in all languages that have words for numbers. Further, numerical irregularity isn't random; rather, it's reliably concentrated in the lower numbers,[5] with the smallest, *one* to *about three or four*, always being the most irregular. As we will see with Mundurukú (Chap. 37), the irregularity of the smallest number-words is a consequence of the way the first words for numbers emerge in any language: as names for familiar objects whose quantity is both small (i.e., perceptible) and dependable. These qualities make such objects useful exemplars of particular quantities, so things that occur naturally in pairs—like the arms, ears, and eyes—might be leveraged to express the idea of *two*. This origin individualizes the words for the smallest numbers to particular societies, languages, and local choices and decisions, so it isn't surprising that the smallest Hindi number-words would be irregular.

In general, however, by the time numbers are being counted beyond *fifty*, their naming has become fairly regular. This tendency reflects a cognitive capacity for building complex higher numbers from smaller simple ones, like *twenty-one* is composed of *two*, *ten*, and *one*. This compositional ability is efficient, for instead of having to memorize the name of every single number we might ever need, we only need to memorize a few names—typically, *one* through *ten* and any higher productive terms like *hundred* and *thousand*—and learn the rules for putting them together. Rule-based composition, in turn, produces words with regular forms, minimizing the need for memorization and often replacing some of the lower (but never the lowest) irregular forms with regular ones.[6] Thus, it's extremely rare for a numerical vocabulary to be irregular beyond the middle decades (multiples of *ten*), since irregularity means that more names must be memorized, and Hindi (Table 25.1) is irregular all the way up to *ninety-nine*.[7]

While sound changes have obscured the pattern, the Hindi numbers are nonetheless formed in a fairly systematic fashion. All the numbers higher than *ten* are composed of a unit (the numbers in the "1–10" column) followed by a decade (the numbers in the "10" row).[8] Thus, each decade plus *one* (e.g., *eleven*, *twenty-one*, *thirty-one*) begins with a version of the word *one* (*ek* ➔ *ik-*), each decade plus *two*

[3] Salomon (1998), p. 10.

[4] Bright (1969); Comrie (2011); McGregor (1987).

[5] Greenberg (1978).

[6] Greenberg (1978).

[7] Bright (1969); Comrie (2011); McGregor (1987).

[8] Bright (1969).

Table 25.1 Hindi number-words one to ninety-nine (adapted from Comrie, 2011, p. 4)

	1–10	11–20	21–30	31–40	41–50	51–60	61–70	71–80	81–90	91–100
1	*ek*	*gyārah*	*ikkīs*	*ikattīs*	*iktālīs*	*ikyāvan*	*iksaṭh*	*ikhattar*	*ikyāsī*	*ikyānve*
2	*do*	*bārah*	*bāīs*	*battīs*	*bayālīs*	*bāvan*	*bāsaṭh*	*bahattar*	*bayāsī*	*bānve*
3	*tīn*	*terah*	*teīs*	*taiṁtīs*	*taiṁtālīs*	*tirpan*	*tirsaṭh*	*tihattar*	*tirāsī*	*tirānve*
4	*cār*	*caudah*	*caubīs*	*cauṁtīs*	*cavālīs*	*cauvan*	*cauṁsaṭh*	*cauhattar*	*caurāsī*	*caurānve*
5	*pāṁc*	*pandrah*	*paccīs*	*paiṁtīs*	*paiṁtālīs*	*pacpan*	*paiṁsaṭh*	*pachattar*	*pacāsī*	*pacānve*
6	*chah*	*solah*	*chabbīs*	*chattīs*	*chiyālīs*	*chappan*	*chiyāsaṭh*	*chihattar*	*chiyāsī*	*chiyānve*
7	*sāt*	*satrah*	*sattāīs*	*saiṁtīs*	*saiṁtālīs*	*sattāvan*	*sarsaṭh*	*sathattar*	*sattāsi*	*sattānve*
8	*āṭh*	*aṭhārah*	*aṭṭāīs*	*aṛtīs*	*aṛtālīs*	*aṭṭhāvan*	*aṛsaṭh*	*aṭhhattar*	*aṭṭhāsī*	*aṭṭhānve*
9	*nau*	*unnīs*	*untīs*	*untālīs*	*uncās*	*unsaṭh*	*unhattar*	*unyāsī*	*navāsī*	*ninyānve*
10	*das*	*bīs*	*tīs*	*cālīs*	*pacās*	*sāṭh*	*sattar*	*assī*	*nave*	*sau*

with a version of *two* (*do* ➔ *ba-* or *bā-*), and so on. Similarly, each decade is also a suffix formed from the words *ten* to *ninety*, analogous to the use of *-teen* for *thirteen* to *nineteen* in English; all the thirties, for example, end in the word *thirty* (*tīs* ➔ *-tīs*), which is itself a version of the word *three* (*tīn*). The formative pattern also systematically changes at the words for *nine*, understood as *one* subtracted from the decade.[9]

In its profound irregularity, Hindi is the polar opposite of Mandarin Chinese (Table 25.2), which is regular throughout. This doesn't mean these number-words originated differently than they did in any other cultural system; rather, it means that when the smaller numbers are combined to produce higher numbers, their pronunciation doesn't change. As was noted for Korean numbers (Chap. 23), transparency is thought to be advantageous in learning and using numbers, since the regularity of the spoken forms makes the numerical structure more apparent.[10]

Some of the region's languages have duodecimal numbers, counting by twelves rather than tens. One is Sora, a language from the eastern portion of India along the Bay of Bengal; another is Chepang, spoken in Nepal, India's Himalayan neighbor to the northeast. Neither is related to each other, as Sora is an Austroasiatic language, and Chepang is Sino-Tibetan.[11] Yet both use an interesting duodecimal type of counting. In Chepang, after counting to *twelve* (*hāle*), *thirteen* is formed as *twelve and one* (*yāt hāle yāt joʔ*), *fourteen* as *twelve and two* (*yāt hāle nis joʔ*), and so on, to *twenty-four* (*two-twelves, nis hāle*), *thirty-six* (*three-twelves, sum hāle pləy.joʔ*), *forty-eight* (*four-twelves, pləy hāle nis joʔ*), and so on.[12] Chepang finger-counting is consistent with these numbers:

> When counting the tip of the thumb is placed against each interstice in turn, starting from the base of the little finger and ending at the tip of the index finger. Since there are four fingers, each with three interstices, this means a total of twelve for each hand and makes 'twelve' a natural base for counting.
>
> (Caughley, 1989, p. 195)

Sora numbers start out in much the same way: after counting to *twelve* (*migel*), *thirteen* is formed as *twelve and one* (*migelboy*), *fourteen* as *twelve and two* (*migelbagu*), and so on, up to *nineteen* (*twelve and seven, migelguiji*).[13] Thereafter, the sequence becomes more interesting. *Twenty*, a "score," breaks the pattern of duodecimal formation, being a unique term (*bokuri*) that suggests vigesimal counting; as is typical in vigesimal counting, *forty* is *two-twenties* (*bakuri*); *sixty*, *three-twenties* (*yakuri*); and *eighty*, *four-twenties* (*unjikuri*),[14] with *three hundred and*

---

[9] Ohala (2002).

[10] Cantlon and Brannon (2007); Fuson and Kwon (1992); Miller et al. (1995); Miller and Stigler (1987); Miura et al. (1988); Miura and Okamoto (1989); Schneider et al. (2020).

[11] Mazaudon (2002).

[12] Caughley (1972, 1989, 1995).

[13] Anderson and Harrison (2008); Mohan (2012).

[14] Anderson and Harrison (2008); Gomango (2020).

Table 25.2 Mandarin number-words *one* to *ninety-nine* (multiple sources)

	1–10	11–20	21–30	31–40	41–50	51–60	61–70	71–80	81–90	91–100
1	*yī*	*shí-yī*	*èr-shí-yī*	*sān-shí-yī*	*sì-shí-yī*	*wǔ-shí-yī*	*liù-shí-yī*	*qī-shí-yī*	*bā-shí-yī*	*jiǔ-shí-yī*
2	*èr*	*shí-èr*	*èr-shí-èr*	*sān-shí-èr*	*sì-shí-èr*	*wǔ-shí-èr*	*liù-shí-èr*	*qī-shí-èr*	*bā-shí-èr*	*jiǔ-shí-èr*
3	*sān*	*shí-sān*	*èr-shí-sān*	*sān-shí-sān*	*sì-shí-sān*	*wǔ-shí-sān*	*liù-shí-sān*	*qī-shí-sān*	*bā-shí-sān*	*jiǔ-shí-sān*
4	*sì*	*shí-sì*	*èr-shí-sì*	*sān-shí-sì*	*sì-shí-sì*	*wǔ-shí-sì*	*liù-shí-sì*	*qī-shí-sì*	*bā-shí-sì*	*jiǔ-shí-sì*
5	*wǔ*	*shí-wǔ*	*èr-shí-wǔ*	*sān-shí-wǔ*	*sì-shí-wǔ*	*wǔ-shí-wǔ*	*liù-shí-wǔ*	*qī-shí-wǔ*	*bā-shí-wǔ*	*jiǔ-shí-wǔ*
6	*liù*	*shí-liù*	*èr-shí-liù*	*sān-shí-liù*	*sì-shí-liù*	*wǔ-shí-liù*	*liù-shí-liù*	*qī-shí-liù*	*bā-shí-liù*	*jiǔ-shí-liù*
7	*qī*	*shí-qī*	*èr-shí-qī*	*sān-shí-qī*	*sì-shí-qī*	*wǔ-shí-qī*	*liù-shí-qī*	*qī-shí-qī*	*bā-shí-qī*	*jiǔ-shí-qī*
8	*bā*	*shí-bā*	*èr-shí-bā*	*sān-shí-bā*	*sì-shí-bā*	*wǔ-shí-bā*	*liù-shí-bā*	*qī-shí-bā*	*bā-shí-bā*	*jiǔ-shí-bā*
9	*jiǔ*	*shí-jiǔ*	*èr-shí-jiǔ*	*sān-shí-jiǔ*	*sì-shí-jiǔ*	*wǔ-shí-jiǔ*	*liù-shí-jiǔ*	*qī-shí-jiǔ*	*bā-shí-jiǔ*	*jiǔ-shí-jiǔ*
10	*shí*	*èr-shí*	*sān-shí*	*sì-shí*	*wǔ-shí*	*liù-shí*	*qī-shí*	*bā-shí*	*jiǔ-shí*	*yī-bǎi*

ninety-nine (*mɨɟgəl-gulɟi-koɽi mɨɟgəl-gulɟi*) being "a score of scores less one."[15] Atypically, the numbers in between the multiples of *twenty* count up from *one* to *twelve*, and then add to *twelve*: *thirty-three* is formed as *twenty and twelve and one* (*bokuri migelboy*), *thirty-four* as *twenty and twelve and two* (*bokuri migelbagu*), and *thirty-nine* as *twenty and twelve and seven* (*bokuri migelgulji*).[16] These forms suggest that Sora was either a duodecimal system that mixed with and borrowed from a vigesimal one, or vice versa. Mixing and borrowing didn't stop there, since the Sora term for *hundred*, *bo-sua*, like the Chepang numbers *five* through *ten*, came from an Indo-European language (unspecified for Sora[17]; Nepali, in the case of Chepang[18]).

References

Anderson, G. D. S., & Harrison, K. D. (2008). Sora. In G. D. S. Anderson (Ed.), *The Munda languages* (pp. 299–380). Routledge.

Bright, W. (1969). Hindi numerals. *Working Papers in Linguistics, 9*, 29–47.

Calkins, P. B., Alam, M., Allchin, F. R., et al. (2024). India. In *Encyclopaedia Britannica*. https://www.britannica.com/place/India

Cantlon, J. F., & Brannon, E. M. (2007). Adding up the effects of cultural experience on the brain. *Trends in Cognitive Sciences, 11*(1), 1–4.

Caughley, R. C. (1972). *A vocabulary of the Chepang language*. Summer Institute of Linguistics.

Caughley, R. C. (1989). Chepang: A Sino-Tibetan language with a duodecimal numeral base? In D. Bradley, E. I. A. Henderson, & M. Mazaudon (Eds.), *Prosodic analysis and Asian linguistics: To honour R. K. Sprigg* (pp. 197–199). Australian National University.

Caughley, R. C. (1995). Chepang, Bagmati province, Nepal. In E. Chan (Ed.), *Numeral systems of the world's languages*. Max Planck Institute. https://lingweb.eva.mpg.de/channumerals/Chepang.htm

Comrie, B. (2011). *Typology of numeral systems*. Trends in Linguistics. Studies and Monographs 118. Max Planck Institute. https://lingweb.eva.mpg.de/channumerals/TypNum_Latest_21ho.pdf

Donegan, P. J., & Stampe, D. (2009). Sora, Rayagada Dt., Orissa, India. In E. Chan (Ed.), *Numeral systems of the world's languages*. Max Planck Institute. https://lingweb.eva.mpg.de/channumerals/Sora.htm

Fuson, K. C., & Kwon, Y. (1992). Korean children's understanding of multidigit addition and subtraction. *Child Development, 63*(2), 491–506.

Gomango, O. (2020). Sora, Rayagada Dt., Orissa, India. In E. Chan (Ed.), *Numeral systems of the world's languages*. Max Planck Institute. https://lingweb.eva.mpg.de/channumerals/Sora.htm

Greenberg, J. H. (1978). Generalizations about numeral systems. In J. H. Greenberg (Ed.), *Universals of human language* (Vol. 3, pp. 249–295). Stanford University Press.

Horo, L., & Sora, D. (2015). Sora, Rayagada Dt., Orissa, India. In E. Chan (Ed.), *Numeral systems of the world's languages*. Max Planck Institute. https://lingweb.eva.mpg.de/channumerals/Sora.htm

[15] Donegan and Stampe (2009).

[16] Anderson and Harrison (2008); Gomango (2020).

[17] Donegan and Stampe (2009); Horo and Sora (2015).

[18] Caughley (1995).

Mazaudon, M. (2002). Les principes de construction du nombre dans les langues tibéto-birmanes [Principles of number construction in Tibeto-Burman languages]. *Mémoires de la Société de Linguistique de Paris, Nouvelle Série, 12*, 91–119.

McGregor, R. S. (1987). *Outline of Hindi grammar with exercises* (2nd ed.). Clarendon Press.

Miller, K. F., Smith, C. M., Zhu, J., & Zhang, H. (1995). Preschool origins of cross-national differences in mathematical competence: The role of number-naming systems. *Psychological Science, 6*(1), 56–60.

Miller, K. F., & Stigler, J. W. (1987). Counting in Chinese: Cultural variation in a basic cognitive skill. *Cognitive Development, 2*(3), 279–305.

Miura, I. T., Kim, C. C., Chang, C.-M., & Okamoto, Y. (1988). Effects of language characteristics on children's cognitive representation of number: Cross-national comparisons. *Child Development, 59*(6), 1445–1450.

Miura, I. T., & Okamoto, Y. (1989). Comparisons of US and Japanese first graders' cognitive representation of number and understanding of place value. *Journal of Educational Psychology, 81*(1), 109–113.

Mohan, S. (2012). Numeral expressions in Kharia, Korku, and Sora: A comparative account. *Bulletin of the Deccan College Research Institute, 72*(73), 367–374.

Ohala, M. (2002). Hindi, India, Nepal, Sri Lanka. In E. Chan (Ed.), *Numeral systems of the world's languages*. Max Planck Institute. https://lingweb.eva.mpg.de/channumerals/Hindi.htm

Salomon, R. (1998). *Indian epigraphy: A guide to the study of inscriptions in Sanskrit, Prakrit, and the other Indo-Aryan languages*. Oxford University Press.

Schneider, R. M., Sullivan, J., Marušič, F., et al. (2020). Do children use language structure to discover the recursive rules of counting? *Cognitive Psychology, 117*, 1–21.

Wright, R. P. (2010). *The ancient Indus: Urbanism, economy, and society*. Cambridge University Press.

Part V
Number Systems in Oceania

Chapter 26
Numbers in New Guinea: Body-Counting

Abstract New Guinea is home to an astonishing number of distinctive numerical systems. Distinctive numbers and counting systems are shibboleths, ways of signaling group affiliation and identity and protecting cultural boundaries. One distinctive method of counting is body-counting, which uses the body as a tally. Body-counting involves the fingers, but in between the two hands, counting proceeds up the arm, up and across the neck and head, and down the other side to create counting cycles like 23 or 27. Before Western education and currency were introduced, body-counting was limited to counting up (or accumulating) an ordinal sequence. Since the body-tally and associated verbal names for the numbers are both ordinal sequences, they are mutually reinforcing in a way that influences the method toward long-term stability. Body-counting often co-exists with one or more alternative methods, including digit-tallying (counting the fingers and/or toes), binary counting (counting by pairs), and quaternary, senary, or duodecimal counting (groups of *four*, *six*, or *twelve*). Together, these methods lack the between-sequence coherence found in the object-specified counting of Polynesia or ancient Sumer.

Keywords Shibboleth · Ordinal sequence · Oksapmin · Yupno · Object-specified counting

After Greenland, New Guinea is the world's largest island. Politically, it is divided in half: the western portion consists of the Indonesian provinces of Papua and West Papua, while the eastern portion is the sovereign nation of Papua New Guinea.[1] With about a fourth of the world's languages (more than 1100), the island is the most linguistically diverse place on the planet, though it is less than one-half of a percent of its surface area.[2] As of 2021, about half of the languages are endangered,

[1] "New Guinea: Island, Malay Archipelago" (2024).

[2] Eberhard et al. (2024); Language Situation in Indonesian Papua (2024).

K. A. Overmann, *Cultural Number Systems*, Interdisciplinary Contributions to Archaeology, https://doi.org/10.1007/978-3-031-83383-0_26

with some having fewer than 10 speakers remaining.[3] The island is also home to an astonishing number of distinctive numerical systems.

Distinctive numbers and counting practices are shibboleths, ways of signaling group affiliation and identity and protecting cultural boundaries against outsiders and external influences. For example, while members of a society count on their hands in the same way as each other, they count on their hands differently or name their fingers differently than their neighbors do, and this helps to maintain a distinctive cultural identity. The manner of counting, or the names used in counting, then identifies the person who counts in a particular fashion as belonging to a particular society or cultural group. As interactions between societies with distinctive counting practices increase and involve numbers, differences tend to disappear over time through the adoption of communal terms and behaviors.[4] Conversely, a high level of numerical distinctiveness among adjacent societies implies that they value social distinctiveness and that perhaps numbers may not be a significant factor in their interaction.

Body-counting uses the body as a tally. In this practice, counting begins on the fingers, as is typical for most of the number systems known throughout the world (the only exceptions don't count as high as *five*). When the fingers are used, counting almost invariably uses the other hand once the first one has reached its full capacity. In body-counting, however, before shifting to the other hand, counting continues up the arm and shoulder, over and across the head, and down the other shoulder and arm. Body-counting tends to be restricted to the upper part of the body, ignoring any parts below the waist, including the toes. The cycles created by body-counting range from a minimum of 12 to a maximum of 74, though 23 and 27 are the most common; about 40 such systems are known, spanning half a dozen language families.[5] While most of these systems are found in the mountainous highlands of New Guinea,[6] one comes from the Torres Strait that separates New Guinea from Australia,[7] and two others (Wotjo and Kulin, Chap. 28) are known in south-east Australia.[8] This distribution suggests a regional practice, though it's also possible "these systems developed independently since there is no evidence of contact between New Guinean Highland peoples and the peoples in South-Eastern Australia."[9]

All these systems differ in some way. Some initiate counting on the right side of the body, others on the left, and a few don't switch from one side to the other.[10] Counting generally begins with the thumb or little finger, and the second hand, once it's reached, can be counted either in the same way or in the reverse of the first hand. As counting proceeds between the two hands, different features of the body are included, and different body features are chosen to act as an unmirrored midpoint

[3] Eberhard et al. (2021).

[4] Saxe (2012).

[5] Lean (1992); Owens (2001).

[6] Comrie (2013).

[7] Ray (1907).

[8] Howitt (1904); Mayer (2018).

[9] Mayer (2018), p. 18.

[10] Franklin and Franklin (1962).

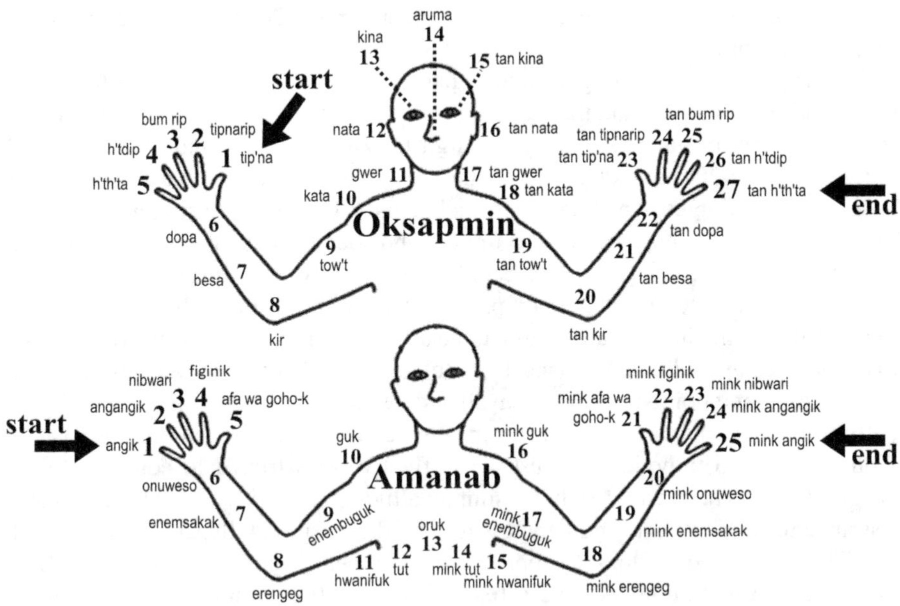

Fig. 26.1 The Oksapmin (top) and Amanab (bottom) body-counting systems. Points on the other side of the body are preceded by a word meaning "other side" (*tan* for Oksapmin; *mink* for Amanab). Top image redrawn and adapted from Saxe (2012, Fig. 12, p. 46); data for bottom image from Lean (1992)

where counting crosses to the other side of the body; when this feature is included, the cycle produces an odd number, an even number if it's not included. One society, the Alamblak, are described as having two body-counting cycles: one is used exclusively by men, the other only by women.[11]

The Oksapmin body-counting system is one of the best known because it has been closely documented.[12] The features distinguishing it from Amanab body-counting are shown in Fig. 26.1. Oksapmin counting sequence starts on the right thumb, proceeds to the right little finger, counts up the right arm (wrist, forearm, inner elbow, and biceps) to the shoulder and neck, counts on the head (ear, eye), reaches the nose as the odd center point, reverses the count down the left side, and finishes by counting from the left thumb to the left little finger.[13] In comparison, the Amanab counting sequence starts with the little finger (the reverse of Oksapmin), proceeds in similar fashion as high as the shoulder, descends to the armpit and counts across the breast, reverses the count from the other shoulder to the wrist, and

[11] Bruce (1984).

[12] Saxe (1981, 2012); also see Lean (1992); Owens (2001).

[13] Lean (1992); Saxe (2012).

counts the fingers in the same order on the other hand to end on the little finger (the same as Oksapmin).[14]

The Yupno body-counting system is particularly interesting. All the fingers on both hands are counted, and then all the toes on both feet, suggesting that the Yupno sequence likely originated as a typical digit-tally system, the kind associated with vigesimal (base-20) numbers.[15] Then, other body parts are added: the earlobes, eyes, nose, nostrils, nipples, navel, testicles, and penis.[16] This is a rare strategy for counting higher than the capacity of the fingers and toes because most other counting systems shift to the digits of other people. Yupno body-counting has other distinctive features: it includes many body parts in the lower half of the body. Anatomical features that come in pairs are counted one after the other, rather than as a sequence that first goes up and then reverses down the other side of the body. It also includes three unmirrored points: the nose, navel, and penis.[17]

Before Western education and currency were introduced, body-counting systems didn't include arithmetic; their use was effectively restricted to counting up, or *accumulating*. The reason for this is simple: although the body might be considered a strange form of tally, it is nonetheless just another form of tally, and as such, it acts just like any other tally does: by representing numbers as an ordinal sequence. In an *ordinal sequence*, numbers have a fixed order, but they don't have any relations between them—maybe not even plus-one, the idea that each number is *one more* than the one before it. In many ways, ordinal numbers are like any other ordinal sequence—for example, they are like the days of the week (Monday, Tuesday, Wednesday, etc.), a sequence of names with order but no numerical relations.

The lack of relations means there's no easy way for ordinal numbers to be added or subtracted, whether materially or mentally. The number-words used in body-counting are also an ordinal sequence, and since the verbal and material forms have the same structure, they are mutually reinforcing; this means the number system is likely to persist unchanged over long periods of time.[18] Further increasing the likelihood of structural persistence, ordinal sequences are stored in memory in a way that suggests the human brain has cognitive resources specialized for and dedicated to them; this cognitive resource is also likely the reason why words for small numbers (*one* through *five*) tend to persist unchanged over very long periods of time.[19]

Body-counting often co-exists with one or more alternative method of counting. For example, in addition to body-counting, the Wiru also practice digit-tally (counting all the fingers and toes) and quaternary counting (counting in groups of *four*); the Ende combine body-counting with digit-tally and senary counting (counting in

---

[14] Lean (1992).

[15] Owens (2001).

[16] Wassmann and Dasen (1994).

[17] Lean (1992); Wassmann and Dasen (1994).

[18] Overmann (2023).

[19] Wynn et al. (2017); see discussion in Overmann (2023).

Table 26.1 New Guinea societies and their counting methods. Compiled from Bowers and Lepi (1975); Evans (2009); Lean (1992); Owens (2001); and Saxe (2012)

Language	Counting methods
Wiru	Body-counting, digit-tally, and quaternary
Ende	Body-counting, digit-tally, and senary
Odoodee	Body-counting and binary
Arapesh	Ternary and quaternary
West Kewa	Body-counting and quaternary
Nambo	Body-counting and senary
Kaninuwa	Digit-tally and duodecimal

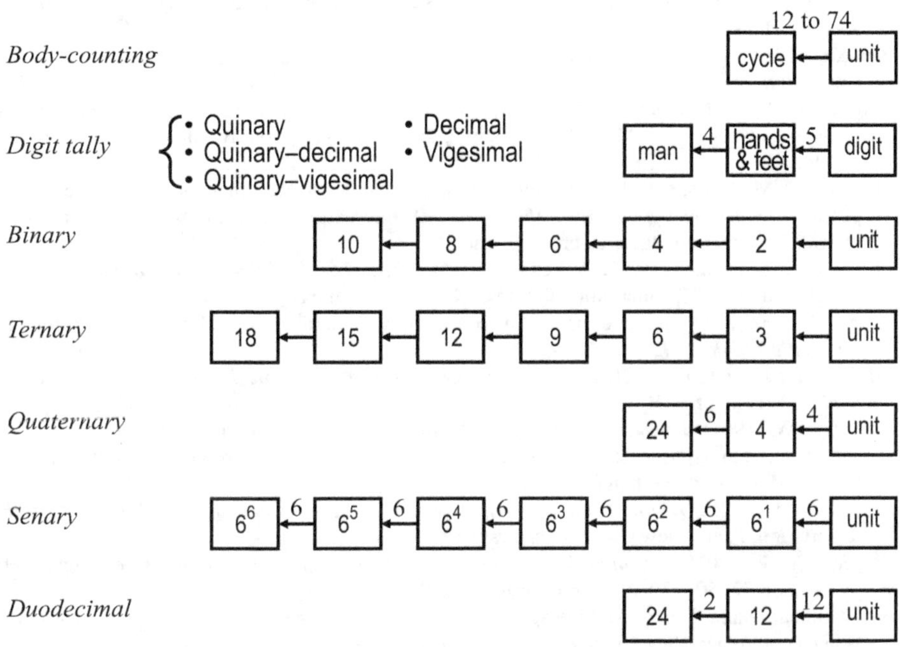

Fig. 26.2 Different methods of counting found in New Guinea. Digit-tally systems can count to *five* (quinary), *ten* (decimal), or *twenty* (vigesimal); the latter two may also retain sub-bases of *five* (quinary-decimal and quinary-vigesimal). Image by the author

groups of *six*); and the Odoodee add binary counting (counting by pairs) to body-counting.[20] It's not unusual for any particular New Guinea society to count using several methods, including duodecimal counting (counting by groups of *twelve*). A selection of New Guinean societies and their counting methods is shown in Table 26.1.

When the different New Guinean counting methods (Fig. 26.2) are considered side-by-side, it is clear that they are object-specified counting. That is, the method of counting used differs according to the object being counted. However, these

[20] Bowers and Lepi (1975); Evans (2009); Lean (1992); Owens (2001); Saxe (2012).

sequences lack the coherence that characterizes the object-specified counting found in Polynesia (Chap. 30) and Sumer (Chap. 2), where the unity between the different sequences suggests that one original counting method was later elaborated into several variants (like counting by singles can give rise to counting by pairs and fours).

References

Bowers, N., & Lepi, P. (1975). Kaugel Valley systems of reckoning. *Journal of the Polynesian Society, 84*(3), 309–324.

Bruce, L. (1984). *The Alamblak language of Papua New Guinea (East Sepik).* Australian National University.

Comrie, B. (2013). Numeral bases. In M. S. Dryer & M. Haspelmath (Eds.), *World atlas of language structures online*. Max Planck Institute for Evolutionary Anthropology. http://wals.info/chapter/131

Eberhard, D. M., Simons, G. F., & Fenning, C. D. (Eds.). (2021). How many languages are endangered? *Ethnologue: Languages of the world.* SIL International. https://www.ethnologue.com/insights/how-many-languages-endangered/

Eberhard, D. M., Simons, G. F., & Fenning, C. D. (Eds.). (2024). *Ethnologue: Languages of the world* (25th ed.). SIL International. https://www.ethnologue.com

Evans, N. (2009). Two *pus* one makes thirteen: Senary numerals in the Morehead-Maro region. *Linguistic Typology, 13*(2), 321–335.

Franklin, K. J., & Franklin, J. (1962). The Kewa counting systems. *Journal of the Polynesian Society, 71*(2), 188–191.

Howitt, A. W. (1904). *The native tribes of south-east Australia.* Macmillan and Co.

Language situation in Indonesian Papua. (2024). *Center for endangered languages documentation.* http://celd.uni-koeln.de/langsit

Lean, G. A. (1992). *Counting systems of Papua New Guinea and Oceania.* Unpublished doctoral dissertation, Papua New Guinea University of Technology.

Mayer, C. J. (2018). *Numeral systems in New Guinea.* https://www.researchgate.net/publication/322506379_Numeral_Systems_in_New_Guinea

New Guinea: Island, Malay Archipelago. (2024). In *Encyclopaedia Britannica.* https://www.britannica.com/place/New-Guinea

Overmann, K. A. (2023). *The materiality of numbers: Emergence and elaboration from prehistory to present.* Cambridge University Press.

Owens, K. (2001). The work of Glendon Lean on the counting systems of Papua New Guinea and Oceania. *Mathematics Education Research Journal, 13*(1), 47–71.

Ray, S. H. (1907). *Reports of the Cambridge Anthropological Expedition to Torres Straits. Vol. 3: Linguistics.* Cambridge University Press.

Saxe, G. B. (1981). Body parts as numerals: A developmental analysis of numeration among the Oksapmin in Papua New Guinea. *Child Development, 52*(1), 306–316.

Saxe, G. B. (2012). *Cultural development of mathematical ideas.* Cambridge University Press.

Wassmann, J., & Dasen, P. R. (1994). Yupno number system and counting. *Journal of Cross-Cultural Psychology, 25*(1), 78–94.

Wynn, T., Overmann, K. A., Coolidge, F. L., & Janulis, K. (2017). Bootstrapping ordinal thinking. In T. Wynn & F. L. Coolidge (Eds.), *Cognitive models in Palaeolithic archaeology* (pp. 197–213). Oxford University Press.

Chapter 27
Numbers in New Guinea: Yam Counting and Other Counting Methods

Abstract A distinctive counting method found in New Guinea counts yams by sorting them into groups of *thirty-six* and then *six*, setting one aside from each group for the next round of counting. This method of counting-by-sorting is similar to one found in Polynesia. Like the Polynesian variant, New Guinean yam-counting creates an *ephemeral abacus*, ephemeral because the abacus exists only while the counting is performed, and an abacus because the method represents exponential structure. The exponential structure created by yam-counting reaches an impressive *six* to the seventh power, or 279,936. Such counting-by-sorting is physically and mentally efficient, minimizing the physical effort of moving items to count them and the mental effort of keeping track of the count. Other distinctive methods found in New Guinea count things by threes (Arapesh), fifteens (Huli), sixteens (Wuvulu), or sixties (Ekari). The Ekari sexagesimal system is the only other one known besides the number system of Mesopotamia. It most likely represents an independent invention, rather than cultural transfer. Iqwaye finger-counting counts to 400 on the fingers and toes of a single person.

Keywords Ephemeral abacus · Exponential structure · Arapesh · Ekari · Iqwaye

Another interesting New Guinean counting method counts yams by sorting them into groups of *thirty-six* and *six*, setting one aside from each group for the next round of counting. This is similar to the method of counting-by-sorting found in Polynesia (Chap. 30). Like Polynesian counting-by-sorting, New Guinean yam-counting creates an "ephemeral abacus": an *abacus* because it creates exponential grouping (tens, hundreds, thousands, etc.) and allows for value to be exchanged between groups; *ephemeral* because the device exists only while counting is being performed.[1] Given the relative proximity of Polynesia and New Guinea, the Polynesian practice may have originated in the western Pacific, and migrated eastward as the Pacific was settled. However, it's also quite common for unrelated

[1] Overmann (2020), p. 69.

K. A. Overmann, *Cultural Number Systems*, Interdisciplinary Contributions to Archaeology, https://doi.org/10.1007/978-3-031-83383-0_27

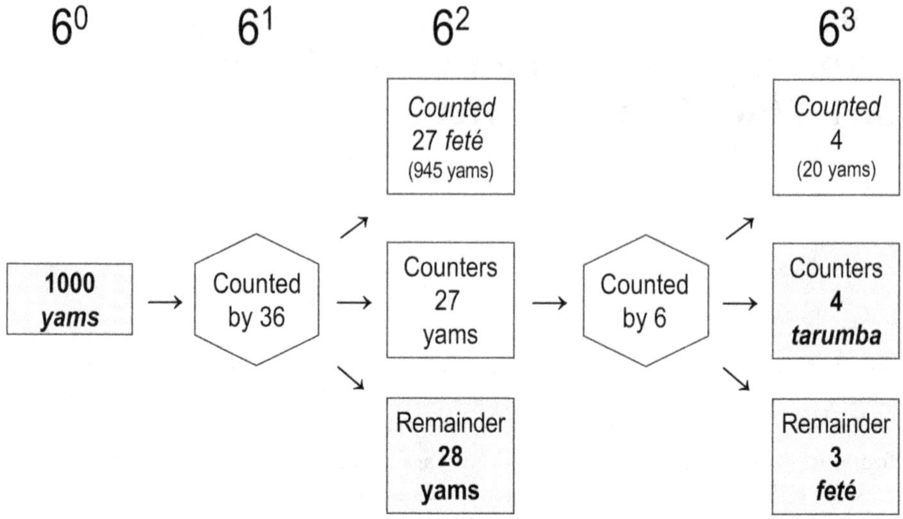

Fig. 27.1 Counting 1000 yams in Arammba. Image by the author based on descriptions in Döhler (2011, 2018), Evans (2009), and Williams (1936)

cultural groups to converge on similar methods, simply because of the shared realities of counting material objects.

Yam-counting counts by six. If we were to count 1000 yams in the Arammba language (Fig. 27.1),[2] the first round creates *feté*, *six* groups of *six* yams. *One* yam from each *feté* is set aside to act as a counter worth *thirty-six* each. These would then be counted in groups of *six*. In total, the process would yield four *tarumba* (worth 216 each), three *feté* (worth *thirty-six* each), and *twenty-eight* loose yams (that is, not enough to make a *feté*).

In Arammba, yam-counting yields an impressive exponential structure, with terms for numbers up to *six* to the seventh power, or 279,936 yams (Fig. 27.2). The highest term, *xàyo xusi*, means "bad smell … an unimaginably large number,"[3] perhaps because there are "too many [yams] to eat so they rot."[4] Similar terms for "uncountable" numbers are found in other languages; the one recorded for the Inka meant a number so high that anyone who wanted to count it would be foolish or disturbed.[5] Ganda (or Luganda), an African language, speaks of any number higher than ten million as "a number which cannot be counted, is like powder, and is not known."[6]

[2] Döhler (2011, 2018); Evans (2009); Williams (1936).

[3] Evans et al. (2018), p. 690.

[4] Boevé-Kroese (2012).

[5] Gonçalez Holguin (1607), Book III, Ch. 52, p. 101.

[6] Atkins (1961), p. 46.

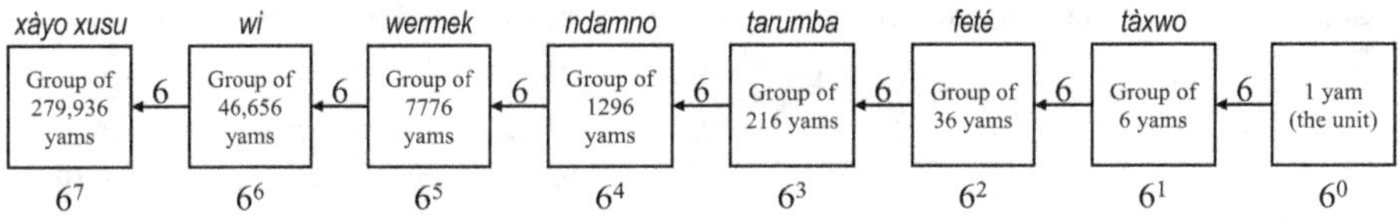

Fig. 27.2 Senary structure in the Arammba language. Image by the author based on data from Boevé-Kroese (2012)

Count things by sorting them is physically and mentally efficient. Humans can't keep track of more than *about three or four* objects just by looking at them, a limitation of both memory and the perceptual system for quantity. Together, these constraints mean that if we want to know how many things there are in a group of objects, we must separate the things we have counted from the things we haven't counted by physically moving them apart from each other. People who aren't used to counting lots of objects often speculate that the way to count is to create groups of, say, *ten*, and then combine them into groups of *hundred*, the hundreds into groups of *thousand*, and so on. This strategy, however, ignores the reality of moving physical objects. For example, an average coconut weighs over 3 pounds, a feasible amount to move; a group of *ten* coconuts is then more than 30 pounds, which is starting to be a lot of weight to shift; a *hundred* nuts weigh over 300 pounds, which is now too much weight for the average person even to consider; and a *thousand* nuts weigh over 3000 pounds, a weight that exceeds even the most phenomenal human strength by a considerable factor.[7] Counting by creating bundles of increasing size is also inefficient because it requires the items to be moved multiple times—once when they are included in a bundle of *ten*, a second time when they are included in a bundle of *hundred*, a third time for a bundle of *thousand*, and so on. So, counting by creating bundles of increasing size isn't a feasible strategy, especially when applied to items numbering in the hundreds, thousands, or more. If this were the method used, no society would ever count much beyond tens of things.

Counting-by-sorting reduces the demands on our physical resources because objects are moved one at a time, though objects that are easily picked up might be counted in groups (counting by pairs, fours, and eights is common). But physical inefficiency isn't the only thing to worry about; there's also mental inefficiency. Consider another strategy that is often proposed by people who aren't used to the practicalities of counting: counting things one at a time. This seems only logical when things are moved one at a time. However, counting sequentially one by one is mentally taxing, especially when numbers go much further than *ten*. It's easy to lose track of the count, especially when lots of things are counted. Simply, the more things are counted, the longer it takes, and the more likely it is to be interrupted and forgotten. Higher numbers are also lengthy and complicated—*one thousand, nine hundred, and seventy-four* is significantly longer and more complex than any of the productive terms that are its components, *one*, *four*, *seven*, *nine*, *ten*, *hundred*, and

[7] Overmann (2021).

Table 27.1 Wuvulu hexadecimal counting system for coconuts (Hafford, 1999, 2014)

Nr.	Wuvulu	Meaning	Nr.	Wuvulu	Meaning
16	*emoro*	1 × 16	256	*'awana'oloroamoro*	16 × 16
32	*ruamoro*	2 × 16	272	*'awana'olorompalomoro*	17 × 16
48	*'olumoro*	3 × 16	288	*'awanafainaroamoro*	18 × 16
64	*runaroamoro*	4 × 16	304	*'awanafaimpalomoro*	19 × 16
80	*aipanimoro*	5 × 16	320	*ruapi'i*	20 × 16
96	*oloroamoro*	6 × 16	160	*epi'i*	10 × 16
112	*'oloroamamoro*	7 × 16	320	*ruapi'i*	20 × 16
128	*fainaroamoro*	8 × 16	480	*'olupi'i*	30 × 16
144	*faimamoro*	9 × 16	640	*runaropi'i*	40 × 16
160	*epi'i*	10 × 16	800	*apanipi'i*	50 × 16
176	*'awanaemoro*	11 × 16	960	*'oloroapi'i*	60 × 16
192	*'awanaruamoro*	12 × 16	1120	*'oloromfuapi'i*	70 × 16
208	*'awana'olumoro*	13 × 16	1280	*fainaroapi'i*	80 × 16
224	*'awanarunaroamoro*	14 × 16	1440	*faimfuapi'i*	90 × 16
240	*'awanaaipanimoro*	15 × 16	1600	*efapu'upi'i*	100 × 16

thousand—and long, complex names not only take more time, they also pose an increase demand on our (already fairly limited) memory. Setting aside every sixth item (or every tenth item, for Polynesian counting-by-sorting) reduces the demands on our mental resources, in two ways: first, only small numbers—from *one* to *six* for New Guinean yam counting; from *one* to *ten* for Polynesian counting—are needed, and second, the items set aside keep track of the count.

The Arapesh count things by threes, a very different method than the Bukiyip counting system we saw in Chap. 3 (Table 3.3). As recounted by the New Zealand anthropologist Reo Franklin Fortune, Arapesh count using multiples of *three*, with a productive term at *six*.[8] When this system is counted on the fingers, the digits are counted, as is the thumb joint, yielding *six* per hand. The term for *both hands* means *twelve*. When the feet are counted in similar fashion, the total reached, *twenty-four*, is called "a man," in parallel with vigesimal counting, where the digits of a person mean *twenty*. An alternative Arapesh method of finger-counting starts with the little, ring, and middle fingers of the first hand as the first unit of *three*; the index finger, thumb, and the thumb of the second hand are counted as the second unit of *three*; the index, middle, and ring fingers of the second hand are counted as the third unit of *three*; and a fourth unit of *three* is counted mentally, as the little finger on the second hand is ignored. Objects counted in threes include betel nuts, big yams, wild game, tobacco braids, hand drums, fire places, moons, garden divisions, and bananas.[9]

The Wuvulu have a hexadecimal system for counting coconuts, used in the early 1900s on coconut plantations. They also have binary and quaternary systems for

[8] Fortune (1942).

[9] Fortune (1942).

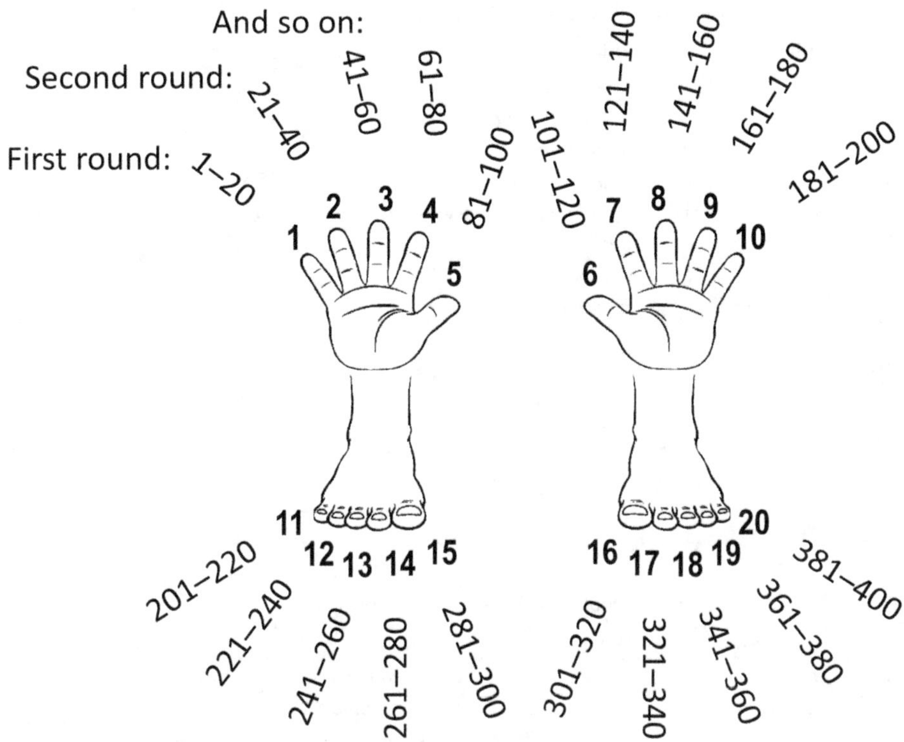

Fig. 27.3 Iqwaye finger-counting. Image by the author based on data from Mimica (1988)

coconuts, and a decimal system for counting that counts into the thousands.[10] The hexadecimal system is shown in Table 27.1.

The Iqwaye count on their fingers to 400. This interesting finger-counting system (Fig. 27.3) begins as vigesimal (base-20) counting. After reaching *twenty* by counting all the fingers and toes on a single person, counting resumes on the same person's digits, which now have a dual purpose: keeping track as counting proceeds from *one* to *twenty* within each cycle, and keeping track of the total number of cycles of *twenty*.[11]

Huli uses fifteen *as a base number.* It's possible this counting system began as body-counting, since *fifteen* wouldn't be an unusual body-counting cycle, and some of the number-words (Table 27.2) do suggest body-parts.[12] However, the parts of the body in question "are not in fact referred to by the Huli when they are counting …

[10] Hafford (1999, 2014).

[11] Mimica (1988).

[12] Cheetham (1978); John Lynch (1998).

Table 27.2 Huli base *fifteen* (Cheetham, 1978, pp. 19, 25)

Nr.	Huli term	Nr.	Huli term	Meaning
1	*mbi*	16	*nguira-ni-mbira*	15 + 1
2	*ki*	18	*nguira-ni-tebira*	15 + 3
3	*tebo*	25	*nguira-ni-pira*	15 + 10
4	*ma*	30	*ngui ki*	15 × 2
5	*dau*	31	*ngui ki ngui tebonego-naga mbira*	30 + one of the third fifteen
6	*waraga*	45	*ngui tebo*	15 × 3
7	*ka*	150	*nguipi*	15 × 10
8	*hali*	225	*ngui ngui*	15 × 15
9	*di*			
10	*pi*			
11	*bea*			
12	*hombe(a)*			
13	*hale*			
14	*de*			
15	*ngui*			

Table 27.3 Ekari base *sixty* (Doble, 1987; Price and Pospisil, 1966)

Nr.	Ekari term	Meaning	Nr.	Ekari term	Meaning
1	*ena*		21	*ena ma mepiina*	1 + 20
2	*wiya*		29	*iyee mepiina*	9 + 20
3	*wido*		30	*yokagaati*	Child of 10
4	*wii*		40	*mepiiya*	2 × 20 (two *mepi*)
5	*idibi*		50	*gaatibeu*	60 – 10 (ten lacking)
6	*benumi*		60	*muto, bado*	60
7	*pituwo*		61	*ena daimita muto*	1 + 60
8	*waguwo*		70	*gaati daimita muto*	10 + 60
9	*iyee*		80	*mepiina daimita muto*	20 + 60
10	*gaati*		100	*mepiiya daimita muto*	40 + 60
11	*ena ma gaati*	1 + 10	[lots]	*muto muto to muto*	Many times 60
12	*wiya ma gaati*	2 + 10			
19	*iye gaati*	9 + 10			
20	*mepiina*	one *mepi*			

and one of [the] elderly informants even went so far as to suggest that any resemblance between the numerals and words for body-parts are purely coincidental!"[13]

Ekari (Kapauku) has a base of sixty. This system (Table 27.3) may also have started as body-counting, though the (highly unlikely and fiercely contested)

[13] Cheetham (1978), p. 16.

possibility that it represents the long-distance migration and survival of Babylonian sexagesimal counting has also been suggested.[14]

References

Atkins, G. (1961). Notes on the concords and classes of Bantu numerals. *African Language Studies, 2*, 42–48.

Boevé-Kroese, A. (2012). Arammba, Papua New Guinea. In E. Chan (Ed.), *Numeral systems of the world's languages*. Max Planck Institute. https://lingweb.eva.mpg.de/channumerals/Arammba.htm

Bowers, N. (1977). Kapauku numeration: Reckoning, racism, scholarship, and Melanesian counting systems. *Journal of the Polynesian Society, 86*(1), 105–116.

Cheetham, B. (1978). Counting and number in Huli. *Papua New Guinea Journal of Education, 14*, 19–30.

Doble, M. (1987). A description of some features of Ekari language structure. *Oceanic Linguistics, 26*(1), 55–113.

Döhler, C. (2011). Kómnzo (Wára), Papua, Indonesia. In E. Chan (Ed.), *Numeral systems of the world's languages*. Max Planck Institute. https://lingweb.eva.mpg.de/channumerals/Wara-Komnjo.htm

Döhler, C. (2018). *A grammar of Komnzo*. Language Science Press.

Evans, N. (2009). Two *pus* one makes thirteen: Senary numerals in the Morehead-Maro region. *Linguistic Typology, 13*(2), 321–335.

Evans, N., Arka, W., Carroll, M., et al. (2018). The languages of southern New Guinea. In B. Palmer (Ed.), *The languages and linguistics of the New Guinea area: A comprehensive guide* (pp. 641–774). De Gruyter Mouton.

Fortune, R. F. (1942). *Arapesh* (Vol. 6). J. J. Augustin.

Gonçalez Holguin, D. (1607) *Gramatica y arte dveva de la lengva general de todo el Peru, llamada lengua Qquichua, o lengua del Inca [Grammar and art of the general language of all Peru, called the Quichua language, or the language of the Inca]*. Francisco del Canto.

Hafford, J. A. (1999). *Elements of Wuvulu grammar*. Unpublished master's thesis, University of Texas at Arlington.

Hafford, J. A. (2014). *Wuvulu grammar and vocabulary*. Unpublished doctoral thesis, University of Hawai'i at Mānoa.

Lynch, J. (1998). *Pacific languages: An introduction*. University of Hawai'i Press.

Mimica, J. (1988). *Intimations of infinity: The mythopoeia of the Iqwaye counting system and number*. Berg.

Overmann, K. A. (2020). The curious idea that Māori once counted by elevens, and the insights it still holds for cross-cultural numerical research. *Journal of the Polynesian Society, 129*(1), 59–84.

Overmann, K. A. (2021). Counting by "elevens" and why *nine* and *two* make *twenty*: The material roots of Polynesian numbers. *Journal of Mathematics and Culture, 15*(3), 1–32.

Price, D. J. d. S., & Pospisil, L. (1966). A survival of Babylonian arithmetic in New Guinea? *Indian Journal of the History of Science, 1*(1), 30–33.

Williams, F. E. (1936). *Papuans of the Trans-Fly*. Clarendon Press.

[14] Bowers (1977); Price and Pospisil (1966).

Chapter 28
Numbers in Australia

Abstract Australia is home to the majority of the world's *restricted* number systems, cultural systems of numbers that count no higher than *twenty*. Three of every four Australian languages count no higher than *three* or *four*. In the late nineteenth century, a few Australian number systems had terms up to *one hundred*. In these systems, numbers higher than *four* show signs of recent emergence in traits like analyzability (etymological transparency showing the use of the hands and feet for counting) and lacking names for numbers between the decades (the names of multiples of *ten* serving as building blocks for the interim numbers). An account from the memoir of William Buckley, an escaped convict who lived with the Wathaurung people for more than three decades, suggests that the lack of names for higher numbers was no barrier to using numerical information, as for example, in a tally of mud-stripes used to keep track of the days in a journey. Timekeeping by means of painting stripes on the body may represent a precursor to body-counting, and several Australian body-counting systems are known. The Gumatj of northern Australia have a quinary (base-5) system.

Keywords Restricted numbers · Wathaurung · Gumatj · Quinary · Analyzability · Etymological transparency

Australia is known for unique flora and fauna, like gum trees (eucalypts) and the koalas that feed exclusively on them. It is also full of deadly plants (nightshade, stinging nettles, and a strychnine tree), animals (venomous snakes and spiders, poisonous jellyfish, man-eating crocodiles, and birds with dagger-like nails), and environmental conditions (deserts, fires, and floods). Humans moved into the continent about 60,000 years ago,[1] and since then, Indigenous people have traded information and goods over thousands of kilometers of challenging terrain.[2] When Western

[1] Clarkson et al. (2017).

[2] O'Conner (2016); Ratsch et al. (2010).

K. A. Overmann, *Cultural Number Systems*, Interdisciplinary Contributions to Archaeology, https://doi.org/10.1007/978-3-031-83383-0_28

contact first began in 1604, about 300 Indigenous languages were spoken;[3] today, there are more than 250, with about 800 dialects.[4] The number systems associated with the Australian languages are known for two qualities: counting to fairly low numbers, and being underestimated in their ability to count.[5]

Australian languages tend to have few terms for numbers. A 2012 sample of 189 Australian languages from thirteen language families and three linguistic isolates showed that most (139, 73.5%) counted no higher than *three* or *four*.[6] Twenty-eight of these languages (14.8%) had forms meaning "hand" for *five* and *ten*,[7] many without the gap numbers *six* through *nine* being filled in. The few counting beyond *ten* are digit-tally counting systems. Examples include Gamilaraay (or Kamilaroi) and Anindilyakwa, both of which reached *twenty* by adding the fingers on both hands and the toes on both feet.[8] As noted in previous chapters, number systems that count no higher than *twenty* are considered to be restricted,[9] and Australia is home to the majority of the world's restricted number systems.

Several Australian languages count to one hundred. In the southeastern region of Victoria, terms up to *one hundred* were recorded in the late nineteenth century for Chaap Wuurong (also called Tyapwurrun or Djab Wurrung) and up to *ninety* for Kuurn kopan noot (Dhauwurd Wurrung).[10] Both were typical vigesimal systems. However, the terms for *eighty*—*puuliit mam baa puuliit mam* in Chaap Wuurong and *puuliit peep baa puuliit peep* in Kuurn kopan not—meant *two twenties and two twenties*, instead of the more typical *four twenties*. Numbers beyond *one hundred* were also expressed in Chaap Wuurong by saying *larbargirrar larbargirrar*, the word for *hundred* twice, while "repeated[ly] opening and shutting the hands"; a similar gesture and expression, *wuurt baa dærang wuurt baa dærang*, expressed numbers larger than *one hundred* in Kuurn kopan noot.[11]

Numbers higher than *four* in these systems show signs of recent emergence. One hallmark of recency is analyzability, and these number-words clearly show the use of the hands and feet for counting. In Chaap Wuurong and Kuurn kopan not, another sign of recency is found in the lack of names for numbers between the decades (e.g., *eleven*, *twenty-two*, *forty-seven*, *ninety-nine*); in both these languages, the intermediate numbers were instead expressed by saying the word for the nearest lower decade (e.g., *ten*, *twenty*, *thirty*) and displaying the requisite number of fingers to be added.[12] This is only what we expect, as number concepts emerge when we engage

[3] Flood (2019).

[4] Languages Alive (2024).

[5] Harris (1982, 1987); Zhou and Bowern (2015).

[6] Bowern and Zentz (2012).

[7] Bowern and Zentz (2012).

[8] Bowern and Zentz (2012); Tindale (1925).

[9] Comrie (2011, 2013).

[10] Bowern and Zentz (2012); Dawson (1881).

[11] Dawson (1881), pp. xcviii–xcix.

[12] Dawson (1881).

material forms, and words for numbers are a secondary phenomenon. Simply, naming occurs when there is something to be named and it starts to recur often enough to be worth the effort of naming and remembering it.

In islands just north of Darwin at the top of Australia, the Tiwi (or Tunuvivi) counted to *ninety-nine* in typical digit-tally fashion, though organized decimally; a term for *hundred* was unknown.[13] As observed, the numbers *one* to *five* are counted on the right hand from thumb to little finger, while *six* through *ten* are counted on the left from thumb to little finger.[14] *Eleven* through *nineteen* are counted by repeating the cycle with the left little finger held down or the left palm struck once (indicating *ten*) with the right index finger, while *twenty* and higher numbers repeat the cycle with the left palm struck the required number of times (i.e., twice for *twenty*, three times for *thirty*, and so on).[15] Since names for the intermediate numbers weren't specified, it's possible that this system too indicated them by combining words and gesture.

The lack of language for numbers is no barrier to using them. An example is found in an account of the life of William Buckley (1780–1856), an Englishman convicted of receiving stolen property (a bolt of cloth) and transported to New South Wales for 14 years or the term of his natural life, whichever came first; he escaped his jailors to live among the Wathaurung people of southeastern Australia for more than three decades.[16] Buckley's memoir is seen as an accurate description of Indigenous Australian life by both contemporary and modern scholars.[17] Early in his memoir, Buckley described a messenger visiting from another tribe to arrange an exchange of eels and roots; the man "had his arms striped with red clay, to denote the number of days it would take us to reach the tribe he came from; … The time stated for this march would be fourteen days."[18]

Several things are interesting about this account. One is the use of numbers—*fourteen* of something—in the absence of any words for numbers higher than *two*. We know the Wathaurung didn't name any numbers beyond *two* because later descriptions are consistent with this restricted limit; one published about 50 years after Buckley's account mentions just *one* and *two*, *ku-i-muty* and *bullaity*,[19] and a second published another 90 years later adds only terms for *three* and *several*, *kulik* and *getjawil*.[20] A second interesting thing is that there really wouldn't be much reason to name the higher numbers associated with the mud stripes because all the messenger needed to do was erase one a day on the return journey, a counterpart to

[13] McRoberts (1990).

[14] McRoberts (1990).

[15] McRoberts (1990).

[16] Morgan (1852).

[17] Curr (1886); Flannery (2017); Tipping (1966).

[18] Morgan (1852), pp. 48–49.

[19] Mathews (1904), p. 732.

[20] Blake et al. (1998), p. 71.

Table 28.1 Wotjo/Kulin body-counting (Howitt, 1904, pp. 697–698)

Nr.	Name	Body-part
1	*giti-munya*	little finger
2	*gaiup-munya*	ring finger
3	*marung-munya*	middle finger
4	*yollop-yollop-munya*	index finger
5	*bap-munya*	thumb
6	*dart-gur*	hollow of radius and wrist
7	*boi-bun*	flexor muscle of forearm
8	*bun-dari*	inside of elbow-joint
9	*gengen-dartchuk*	upper arm
10	*borporung*	point of shoulder
11	*jarak-gourn*	neck
12	*nerup-wrembul*	base of ear
13	*wurt-wrembuly*	head above and just behind ear
14	*doke-doke*	muscle seen when eating
15	*det-det*	crown of head
16–29	[14 to 1, the reverse order]	[14 to 1, the reverse order]

adding one a day on the inbound trip.[21] No number-words are needed because the days wouldn't need to be individually numbered or individuated in any way. Any particular stripe would be numerically uninteresting, indicating only another day in the journey, rather than meaning, for example, the third or fifth day of travel. As has been noted, "What is culturally absent from some but not all Aboriginal groups is not so much the need to count but the need to count precisely."[22]

Third, it's also possible the Wathaurung mud stripes were a precursor to body-counting, which in turn might represent a logical progression from timekeeping by means of painting stripes on the body. One such body-counting system was described for the Wotjo (or Wotjobaluk) and Kulin nations in the southeastern Victoria region.[23]

Australian body-counting (Table 28.1) resembles the body-counting systems found in New Guinea (Chap. 26). Like the New Guinean systems, the Australian one uses the body as a tally, marks it with clay, and uses the marks to keep track of days, either as a proxy for the distance traveled during a journey or as the time to elapse before an event.[24] Australian body-counting also starts and ends with the fingers, moves up the arm to the head, and has an odd (unmirrored) center point, after which counting descends the body in the reverse order. In addition, it counts to 29, producing a cycle similar in its extent to the New Guinean systems. The names for the body-parts counted in reverse order would most likely have included a term like "second" or "other," though one wasn't specified in the source document.

[21] Howitt (1904).

[22] Harris (1982), p. 166.

[23] Howitt (1904).

[24] Mayer (2018).

Despite their resemblances, the two systems are not identical, suggesting they may have developed independently.[25] A key difference is that in New Guinea, the names of the positions tend to be those of the body-parts themselves, whereas in the Australian system, the names of the positions may be things the body-parts look like; for example, the middle finger is named after *marung*, the desert pine (*Callitris verucosa*), a tree known for its height.[26] Some of the names in the Australian system also refer to ornaments worn on the body, rather than the body itself: the upper arm is "the place where the armlet of opossum pelt is tied round the biceps," while the neck is "the place where the reed necklace is worn."[27]

The Gumatj number system of Northern Australia counts by fives. These quinary numbers have an explicit exponential structure: *five* ("10" in base-5, or 10_5) is *wang-gany rulu*, or *one [times] five*; *twenty-five* (100_5) is *dambumirri rulu*, or *[one] five-squared*; *one hundred and twenty-five* (1000_5) is *dambumirri dambumirri rulu*, or *[one] five-cubed*; and *six hundred and twenty-five* ($10{,}000_5$) is *marrma dambumirri dambumirri dambumirri rulu*, or *[one] five(five-cubed)*.[28] In English, a counterpart would be *ten tens* for *hundred*, the term in English's ancestor, Proto-Indo-European;[29] a similar term for *thousand* would be the hypothetical *ten(ten-tens)*. While the Gumatj who provided the numbers may have counted much higher for the observer than they typically might in daily life,[30] they certainly showed they understood how exponents work.

References

Blake, B. J., Clark, I. D., & Krishna-Pillay, S. H. (1998). Wathawurrung and the Colac language of Southern Victoria. In B. J. Blake (Ed.), *Wathawurrung: The language of the Geelong–Ballarat area* (pp. 59–154). Pacific Linguistics.

Bomhard, A. R. (2008). Some thoughts on the Proto-Indo-European cardinal numbers. In J. D. Bengtson (Ed.), *In hot pursuit of language in prehistory. Essays in the four fields of anthropology in honor of Harold Crane Fleming* (pp. 213–222). John Benjamins.

Bowern, C., & Zentz, J. (2012). Diversity in the numeral systems of Australian languages. *Anthropological Linguistics, 54*(2), 133–160.

Clarkson, C., Jacobs, Z., Marwick, B., et al. (2017). Human occupation of northern Australia by 65,000 years ago. *Nature, 547*(7663), 306–310.

Comrie, B. (2011). *Typology of numeral systems*. Trends in Linguistics. Studies and Monographs 118. Max Planck Institute. https://lingweb.eva.mpg.de/channumerals/TypNum_Latest_21ho.pdf

Comrie, B. (2013). Numeral bases. In M. S. Dryer & M. Haspelmath (Eds.), *World atlas of language structures online*. Max Planck Institute for Evolutionary Anthropology. http://wals.info/chapter/131

[25] Mayer (2018).

[26] Howitt (1904).

[27] Howitt (1904), pp. 697–698.

[28] Harris (1982), pp. 170–171.

[29] Bomhard (2008).

[30] Harris (1982).

Curr, E. M. (1886). *The Australian race: Its origin, languages, customs, place of landing in Australia, and the routes by which it spread itself over that continent* (Vol. 2). John Ferres, Government Printer.

Dawson, J. (1881). *Australian Aborigines: The languages and customs of several tribes of Aborigines in the western district of Victoria, Australia.* George Robertson.

Flannery, T. (2017). Introduction. In T. Flannery (Ed.), *The life and adventures of William Buckley: Text classics* (pp. 1–12). Text Publishing.

Flood, J. (2019). *The original Australians: The story of the Aboriginal people* (2nd ed.). Allen & Unwin.

Harris, J. (1982). Facts and fallacies of Aboriginal number systems. *Works Papers of SIL-AAB, Series B, 8*, 153–181.

Harris, J. (1987). Australian Aboriginal and Islander mathematics. *Australian Aboriginal Studies, 2*, 29–37.

Howitt, A. W. (1904). *The native tribes of south-east Australia.* Macmillan and Co.

Languages alive. (2024). *Australian Institute of Aboriginal and Torres Strait Islander Studies (AIATSIS).* https://aiatsis.gov.au/explore/languages-alive

Mathews, R. H. (1904). Language of the Wuddyāwūrru Tribe, Victoria. *Zeitschrift für Ethnologie, 36*(6), 729–734.

Mayer, C. J. (2018). *Numeral systems in New Guinea.* https://www.researchgate.net/publication/322506379_Numeral_Systems_in_New_Guinea

McRoberts, R. W. (1990). Counting at Pularumpi: A survey of a traditional mathematics and its implications for modern learning. *Aboriginal Child at School, 18*(2), 19–43.

Morgan, J. (1852). *The life and adventures of William Buckley: Thirty-two years a wanderer amongst the Aborigines of the then unexplored country around Port Phillip, now the Province of Victoria* (19th ed.). Griffin Press.

O'Conner, M. R. (2016). *Wayfinding: The science and mystery of how humans navigate the world.* St. Martin's Press.

Ratsch, A., Steadman, K. J., & Bogossian, F. (2010). The *pituri* story: A review of the historical literature surrounding traditional Australian Aboriginal use of nicotine in Central Australia. *Journal of Ethnobiology and Ethnomedicine, 6*(1), 1–13.

Tindale, N. B. (1925). Natives of Groote Eylandt and of the west coast of the Gulf of Carpentaria, Part 2. *Records of the South Australian Museum, 3*(2), 103–134.

Tipping, M. J. (1966). Buckley, William (1780–1856). *Australian dictionary of biography.* https://adb.anu.edu.au/biography/buckley-william-1844/text2133

Zhou, K., & Bowern, C. (2015). Quantifying uncertainty in the phylogenetics of Australian numeral systems. *Proceedings of the Royal Society B: Biological Sciences, 282*(1815), 20151278.

Chapter 29
Numbers in Melanesia and Micronesia

Abstract Melanesia is the region of the Pacific that includes the islands from New Guinea to Fiji. Micronesia consists of the islands to the north of Melanesia. The counting practices of Melanesia and Micronesia include counting by pairs. This method is similar to the pair-counting found in Polynesia, in that pairs are counted from *one* to *ten*, and the total count is understood as doubled. Collaborative counting, a method of counting found in Polynesia in which different people keep track of the units, tens, hundreds, and thousands, is also found in Fiji at Melanesia's eastern border with Polynesia. In the eastern part of Micronesia, Marshall Islanders used stones to tally every ten coconuts. This practice created an external record, which is advantageous in that the external record (the collected stones) can be used away from the goods being counted (the coconuts). It is also a disadvantage because collecting, carrying, and counting stones involves effort that counting-by-sorting does not. In the western part of Micronesia, the Chamorro language of Guam shows dramatic change in its vocabulary of number-words from several centuries of Spanish influence.

Keywords Pair-counting · Collaborative counting · Guam · Fiji · Language change

Melanesia is the region of the Pacific that includes the islands from New Guinea to Fiji, while Micronesia consists of the islands to its north. Both were named in 1832 by the French explorer and naval officer Jules Dumont d'Urville, Melanesia for the color of its inhabitants—"la race noire océanienne"—and Micronesia for its very small islands, "îles très-petites"[1] (the name Polynesia, meaning "multiplicity of islands" and originally encompassing all of the islands in the Pacific, had been created in 1756 by the French scholar Charles de Brosses[2]). Beyond his role in naming the two regions, D'Urville is notable for being the second-in-command of the

[1] D'Urville (1832), pp. 5–6.

[2] De Brosses (1756), p. 80.

K. A. Overmann, *Cultural Number Systems*, Interdisciplinary Contributions to Archaeology, https://doi.org/10.1007/978-3-031-83383-0_29

Table 29.1 Pair-counting in the Duke of York islands, Fiji, and Samoa (Brown, 1877, p. 147; Codrington, 1885, pp. 571–572)

Nr.	Duke of York	Fiji	Samoa	Meaning
1	*te kai*	*e dua*	*e tasi*	One pair = two
2	*u rua*	*e rua*	*e lua*	Two pairs = four
3	*u tul*	*e tolu*	*e tolu*	Three pairs = six
4	*lu wal*	*e va*	*e fa*	Four pairs = eight
5	*ti lim*	*e lima*	*e lima*	Five pairs = ten
6	*ma nom*	*e ono*	*e ono*	Six pairs = twelve
7	*ma wil*	*e vitu*	*e fitu*	Seven pairs = fourteen
8	*ti wal*	*e walu*	*e valu*	Eight pairs = sixteen
9	*ti wur*	*e ciwa*	*e iva*	Nine pairs = eighteen
10	*ti kino*	*e tini*	*e sefulu*	Ten pairs = twenty

Coquille during her 1822–1825 circumnavigation of the globe, a voyage that would become associated with one of the more famous and persistent hoaxes associated with Polynesia, the idea that Māori counted by elevens (Chap. 31). D'Urville was in command of the same ship on her next voyage (1826–1829). By then, the *Coquille* had been renamed the *Astrolabe* in honor of one of the two ships of the *La Pérouse* expedition, both lost in 1788. D'Urville confirmed the location of the remains of the namesake ship off the coast of Vanikoro, which lies east of the Solomon Islands and north of Vanuatu; the second *La Pérouse* ship wouldn't be found until 1964. An unsuccessful 1791–1794 expedition sent in search of the same ships had returned to France emptyhanded, save for a spurious and profane vocabulary of Tongan number-words (also covered in Chap. 31).

Melanesia provides insight into regional counting practices. Counting by pairs, which is quite common in traditional Polynesian counting (Chap. 30), is found throughout Melanesia, from its westernmost extent to its easternmost. For example, pair counting (Table 29.1) was attested in the Duke of York islands in the Bismarck Archipelago, which lies off the northeastern coast of New Guinea, as well as Fiji, which lies just west of Polynesia and is often used as a closely related comparison for Polynesian languages. As was true in Polynesian pair-counting, Melanesian pair-counting counted pairs from *one* to *ten*, and the total count was understood as doubled.[3]

In another example of regional relatedness, the term most commonly used for *hundred* in Melanesia and Polynesia is *rau* (or *lau*, a shift in pronunciation in Polynesia), meaning *leaf* or *branch*. In the Vanuatu islands to the northwest of Fiji, palm-like cycas fronds known as *mele* were used in counting: "To count the days after a death a *mele* frond was taken, and beginning on one side of it a leaflet was counted for each day, one being pinched down as a tally for every tenth. The frond when treated in this way on both sides furnished tallies for a hundred, and the final

[3] Brown (1877).

Table 29.2 Fijian numbers (Hazlewood, 1850)

Nr.	Power	Fijian	Nr.	Power	Fijian
1000	10^3	*udolu*	1,000,000	10^6	*vetelei* or *oka niu*
10,000	10^4	*oba*	10,000,000	10^7	*wakaniu*
100,000	10^5	*vatuloa*			

death-feast was commonly held on the hundredth day."[4] Cycas fronds were also used when a dying man's affairs were put in order: "He sends out cycas leaf tallies, on which the fronds on the left-hand side are plucked off so as to indicate his debts, and those on the right-hand the number and grade of pigs which he is owed.[5]

In Fiji, we also find a term, *bunu-ca*, that means the practice of counting collaboratively, in which one person counts the units, another keeps track of the tens, a third the hundreds, and so on.[6] This suggests the collaborative counting-by-sorting used for yams in New Guinea was related culturally to similar forms found throughout Polynesia, including the collaborative counting found in Tonga[7] and Tikopia[8] in western Polynesia and the counting-by-sorting found in central and eastern Polynesia (discussed in detail in Chap. 30). Just like Polynesian counting did at the time of European contact, Fijian counting similarly extended to very high numbers, with term for *million* and *ten million* (Table 29.2). Beyond this upper limit, there was a term for "count[ing] many [things] but uselessly," *vakasaokaoka walega.*[9]

The extent of Fijian counting is comparable to the heights reached in Polynesia and Micronesia. In Polynesia, the highest extent was found in the Marquesas (*ten million*, or *ten* to the seventh power).[10] In Micronesia, the highest extents were found in Pohnpei (a *billion*, or *ten* to the ninth power) and Nukuoro (a most impressive *ten billion*, or *ten* to the tenth power).[11] Of course, it's unlikely people were counting things like fish or coconuts to such great numbers, raising the question of what they would have been used for: "While it is undoubtedly true that there were no objects in any society that one would want to count into the hundreds of thousands or millions, we do not find it inconceivable that an abstract mathematics with linguistic means of representing quantities of such magnitude or beyond could have developed in traditional Oceanic societies."[12]

In eastern Micronesia, stones were used in tallying. In counting coconuts in the Marshall Islands, a stone was used as a mark for every ten coconuts ("er legt für

[4] Codrington (1885), p. 249.
[5] Layard (1942), p. 537.
[6] Hazlewood (1850), p. 20.
[7] Collocott (1925, 1927).
[8] Durrad (1913).
[9] Hazlewood (1850), p. 98.
[10] Dordillon (1857); Hale (1846).
[11] Harrison & Jackson (1984).
[12] Harrison & Jackson (1984), p. 61.

Table 29.3 Palauan and Chamorro number words (Keate, 1788, pp. 377–378; Von Chamisso, 1821, pp. 412–413)

Nr.	Chamorro	Meaning	Palau	Meaning
1	*hatjijai*		*tong*	
2	*huguijai*		*oroo*	
3	*totguijai*		*othey*	
4	*fatfatái*		*oang*	
5	*limijai*		*aeem*	
6	*gonmaijai*		*malong*	
7	*fedguijai*		*oweth*	
8	*gnalguijai*		*tei*	
9	*siguijai*		*etew*	
10	*munutai*		*makoth*	
20	*huguanafulu*	10 × 2	*oloyuck*	2 × 10
30	*tulungafulu*	10 × 3	*ockatbey*	3 × 10
40	*fatfatnafulu*	10 × 4	*ockawaugh*	4 × 10
50	*limangafulu*	10 × 5	*ockeem*	5 × 10
60	*gonumnafulu*	10 × 6	*ockgollom*	6 × 10
70	*fitinafulu*	10 × 7	*ockgweeth*	7 × 10
80	*gualungafulu*	10 × 8	*ockeye*	8 × 10
90	*siguanafulu*	10 × 9	*ockatuew*	9 × 10
100	*gatus*		*mackoth adart*	10 × 10
1000	*tjalan*			

zehn Kokosnüsse einen Stein als Marke"[13]). Compared to the counting-by-sorting found in New Guinea and Polynesia and the collaborative counting found in Fiji and Tonga, this type of counting has the advantage of creating a record external to the goods being counted. It also has a drawback, creating a record external to the goods being counted. Yes, the same thing is both advantageous and disadvantageous. It's advantageous because the external record (the collection of stones) persists once the goods (coconuts, in this case) have been counted, and because the external record can be used away from the enumerated goods. It's disadvantageous because creating and using the external record are tasks, in and of themselves: the stones have to be collected, carried, and counted, and the amount of effort involved rises with the number of stones.

Counting by pairs was also practiced in the western part of Micronesia: in Palau, when "counting out *fish*, they proceed by pairs or couples."[14] Palau (or Pelew, the archaic spelling) was also known at the time as Lord North Island, named after Frederick North, Prime Minister of England during the American Revolutionary War.

Vocabularies of number-words (Table 29.3) were collected for the islands of Palau in 1788 and Guam in 1821. As an indication of language change, consider that

[13] Krämer (1906), p. 438.

[14] Pickering (1845), p. 240.

in 1565, the Spanish navigator Esteban Rodríguez recorded the Chamorro numbers up to *ten* as *acha*, *gua*, *tero*, *farfur*, *nimi*, *guanam*, *frintin*, *gua*, *agua*, and *manete*,[15] and in 1668, the Spanish missionary Diego Luis de San Vitores listed the Chamorro numbers as *acha*, *ugua*, *tiro*, *fafat*, *rima*, *gunum*, *piti*, *uaru*, *sigua*, *monot*, *monot nagay acha* (*eleven*), *monot nagay ugua* (*twelve*), *ugua naporo* (*twenty*), *tiro naporo* (*thirty*), *acha nasarum* (*one hundred*), *ugua nasaran* (*two hundred*), *tiro nasaran* (*three hundred*), and *monot nasaran* (*ten hundred*) or *gatus* (*thousand*).[16]

References

Brown, G. (1877). Notes on the Duke of York Group, New Britain, and New Ireland. *Journal of the Royal Geographical Society, 47*, 137–150.

Burrus, E. J. (1954). Sanvitores' grammar and catechism in the Mariana (or Chamorro) language (1668). *Anthropos, 149*(5/6), 934–960.

Codrington, R. H. (1885). *The Melanesian languages*. Clarendon Press.

Collocott, E. E. V. (1925). Supplementary Tongan vocabulary; also notes on measuring and counting, proverbial expressions and phases of the moon. *Journal of the Polynesian Society, 34*(3), 193–213.

Collocott, E. E. V. (1927). Kava ceremonial in Tonga. *Journal of the Polynesian Society, 36*(1), 21–47.

D'Urville, J. D. (1832). Sur les îles du grand ocean [On the islands of the great ocean]. *Bulletin de la Société de Géographie, 105*(1), 1–21.

De Brosses, C. (1756). *Histoire des navigations aux terres australes [History of navigation in the southern lands]* (Vol. 1).

Dordillon, R. I. (1857). *Essai de grammaire de la langue des Îles Marquises [Essay on the grammar of the language of the Marquesas Islands]*. Imprimerie du Commerce.

Durrad, W. J. (1913). A Tikopia vocabulary. *Journal of the Polynesian Society, 22*(86), 86–95.

Hale, H. (1846). *United States exploring expedition. During the years 1838, 1839, 1840, 1841, 1842. Under the command of Charles Wilkes, U.S.N. Vol. IV: Ethnography and philology*. C. Sherman.

Harrison, S. P., & Jackson, F. H. (1984). Higher numerals in several Micronesian languages. In B. W. Bender (Ed.), *Studies in Micronesian linguistics* (Vol. C-80, pp. 61–79). Australian National University.

Hazlewood, D. (1850). *A Feejeean and English dictionary: With examples of common and peculiar modes of expression, and uses of words*. Wesleyan Mission Press.

Keate, G. (1788). A vocabulary of the Pelew language. In *An account of the Pelew Islands, situated in the western part of the Pacific Ocean. Composed from the journals and communications of Captain Henry Wilson and some of his officers, who, in August 1783, were there shipwrecked, in the Antelope* (pp. 365–378). Luke White.

Krämer, A. (1906). *Hawaii, Ostmikronesien und Samoa: Meine zweite Südseereise (1897–1899) zum Studium der Atolle und ihrer Bewohner [Hawaii, Eastern Micronesia and Samoa: My second South Sea voyage (1897–1899) to study the atolls and their inhabitants]*. Strecker & Schröder.

Layard, J. (1942). *Stone men of Malekula: Vao*. Chatto & Windus.

15 Rodríguez-Ponga (2013).

16 Burrus (1954).

Pickering, J. (1845). *Memoir on the language and inhabitants of Lord North's Island.* Metcalf and Company.

Rodríguez-Ponga, R. (2013). Esteban Rodríguez' vocabulary of the language of Guam (1565). In S. R. Fischer (Ed.), *Oceanic voices–European quills: The early documents on and in Chamorro and Rapanui* (pp. 25–52). Akademie Verlag GmbH.

Von Chamisso, A. (1821). Vocabulary of the dialects Chamori (Mariana Islands), and of Eap, Ulea, and Radack. In O. Von Kotzebue (Ed.), *A voyage of discovery, into the South Sea and Beering's Straits, for the purpose of exploring a north-east passage, undertaken in the years 1815–1818, at the expense of his Highness the Chancellor of the Empire, Count Romanzoff, in the ship* Rurick (Vol. II, pp. 411–433). Longman, Hurst, Rees, Orme, and Brown.

Chapter 30
Numbers in Polynesia

Abstract Polynesians traditionally counted by sorting, setting aside every tenth item to represent *ten* of the items being counted. Large items were counted singly and smaller items by pairs and fours, creating side-by-side counting sequences or "object-specified counting." Counting-by-sorting minimized the physical effort of moving items to count them and the mental effort of keeping track of the count. The method explains why *iwakalua*, the Hawaiian word for *twenty*, means *nine and two*: in counting pairs, every *nine* pairs (18 items) are sorted and the tenth pair (*two* items) is set aside as a counter. Polynesians also had a collaborative method where one person counted the units, a second person counted the tens, a third person the hundreds, and so on. These methods created an *ephemeral abacus*, ephemeral because the abacus existed only while counting was performed and an abacus because the method represented exponential values. In Mangareva in eastern Polynesia, a unique binary variant of counting-by-sorting developed. This method leveraged the binary relations created by counting by singles, pairs, fours, and eights; the result anticipated Western binary computation by centuries.

Keywords Counting-by-sorting · Collaborative counting · Ephemeral abacus · Mangareva · Object-specified counting

Polynesia is a triangular region of the Pacific Ocean bounded by Hawai'i to the north, Rapa Nui (Easter Island) to the east, and Aotearoa (New Zealand) to the south. It contains over a thousand islands, mostly volcanic but some created by continental submergence. The migration that colonized the region over the last three millennia originated in Taiwan and, ultimately, China.[1] Beyond its western border, Polynesia developed in relative isolation,[2] making it an ideal "laboratory" for studying numbers, especially in its peripheries.[3] The traditional number system counted

[1] Chambers (2013); Chambers and Edinur (2020); Guérin (2017); Hurles et al. (2003); Lipson et al. (2018).

[2] Kirch (1984), p. 3.

[3] Suggs (1961), p. 194.

K. A. Overmann, *Cultural Number Systems*, Interdisciplinary Contributions to Archaeology, https://doi.org/10.1007/978-3-031-83383-0_30

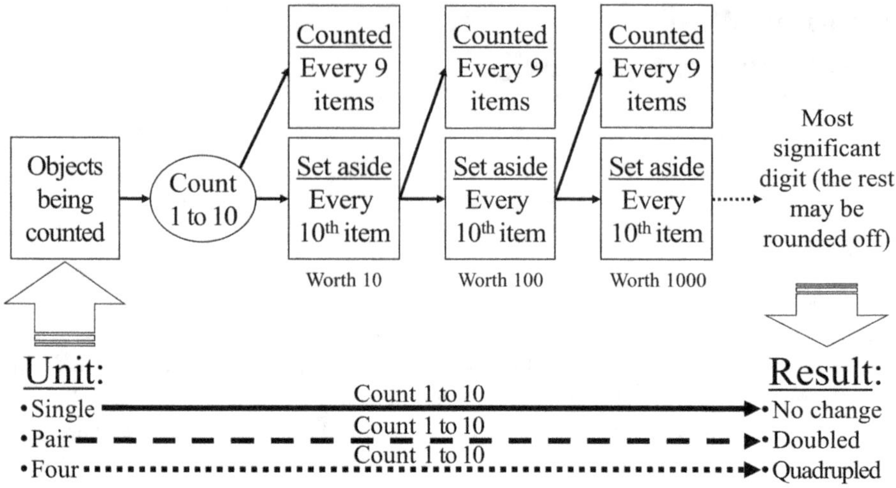

Fig. 30.1 Polynesian counting-by-sorting. Image by the author

into the tens of thousands—perhaps even into the millions—when Western explorers first encountered it in the early seventeenth century.[4] One of its most unusual features was its method of counting.

Polynesians counted by sorting. In this method, every tenth item was set aside to represent *ten* of the items being counted[5] (Fig. 30.1). Then, if there were more than ten of them, the items set aside were counted in the same way, with every tenth item now representing a *hundred*. In the third round, every tenth item would represent a *thousand*, and so on. Counting continued until fewer than ten items remained, representing the most significant digit. For example, if three items remained after three rounds of counting, it meant that 3000 items had been counted. Although the remainders of the previous rounds indicated the other digits, these were typically dismissed with terms that meant "remainder."

Polynesians counted different types of objects differently. Things that were too large to be picked up and sorted, like people and canoes, were counted singly, while smaller, more portable items like coconuts and fish were counted in pairs or groups of four. The process itself didn't vary—counting still proceeded from *one* to *ten*, with every tenth unit set aside—but the result was understood as doubled when the unit was a pair, quadrupled when the unit was a group of four. So, if three units remained after three rounds of counting, 6000 items had been counted if the unit was a pair, 12,000 items with groups of four.

Counting by singles, pairs, and fours as the base unit created side-by-side counting sequences, or "object-specified counting." We get some sense of this when we think about how we count by singles, *one-two-three-four*; by twos, *two-four-six-eight*; or by

[4] Cook (1777); Roggeveen (1721).

[5] Overmann (2020).

fives, *five-ten-fifteen-twenty*. We think of numerical values as proceeding along a single continuum or number line, where the interval between numbers is always *one*, even if we skip along by twos or fives. In contrast, the Polynesian sequences counted from *one* to *ten* while varying the base unit. This created side-by-side counting sequences that were united by their equivalences: *twenty* singles, *two* tens, *ten* pairs, and *five* groups of four.

The method of counting-by-sorting resolves a historical mystery. The Hawaiian word for *twenty*, *iwakalua*, means *nine and two*. This has puzzled observers ever since it was first reported almost two centuries ago[6] because it doesn't add up—quite literally. Number-words are usually quite commonsensical in their arithmetical formations, but in this case, regardless of how they are added or multiplied, 9 and 2 simply don't make 20. However, once we understand that some commodities were picked up in pairs, that pairs were counted from *one* to *ten*, that the count was understood as doubled, and that the last pair was treated differently in being set aside to act as a counter, then *iwakalua* makes perfect sense as a term for *twenty*. It is *nine* pairs sorted into one pile, and the last pair, *two* items, set aside for the next round of counting.[7]

Counting by pairs and fours was not unique to Polynesia, as traces of the practice are also found in European languages:

> The earliest and simplest mode of reckoning was by *pairs*, arising naturally from the circumstance of both hands being employed in it, for the sake of expedition. It is now familiar among sportsmen, who use the names of *brace* and *couple*, words that signify *pairing* or *yoking*. … The numbering by *fours* … was evidently suggested by the custom of taking, in the rapid tale of objects, a pair in each hand. Our fishermen, who generally count in this way, call every double pair of herrings, for instance, a *throw* or *cast*; and the term *warp*, which, from its German origin, has exactly the same import, is employed to denote *four*, in various articles of trade.
>
> (Leslie, 1820, p. 2, emphasis in original)

Traces of counting by singles, pairs, and fours have also been reported for Khmer, an Austroasiatic language of Cambodia (Fig. 30.2).[8] If the Khmer practice represented a cultural tradition shared with Polynesia, it would be thousands of years old.[9]

The practice of counting by pairs and fours has greatly confused Western observers. This suggests that any such practice in Western societies has long since vanished from our collective memory. For the most part, this confusion has meant that Polynesian number systems are often misclassified, generally as vigesimal (base-20), when in actuality they are straightforwardly decimal (base-10) in their organization. Perhaps the earliest incident of this sort occurred in 1821, when the German botanist Adelbert von Chamisso, confused by the practice of pair-counting, published

[6] Andrews (1854); Clark (1839); Von Chamisso (1837).

[7] Overmann (2020); Wynn et al. (2021).

[8] Jenner (1976); also see Aymonier (1883); Cœdès (1924).

[9] Chambers (2013); Chambers and Edinur (2020).

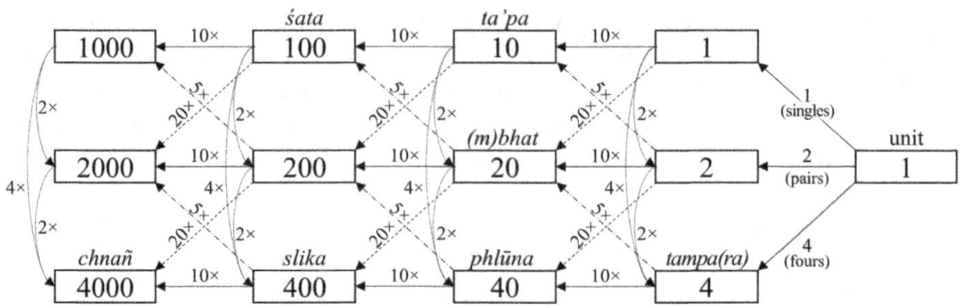

Fig. 30.2 Khmer counting by singles, pairs, and fours. Image by the author

work that identified the New Zealand number system as based on *twenty*.[10] He would publish a correction in just a few years,[11] quite a fast pace, given the slower pace of traveling, publishing, and corresponding in the early nineteenth century. In Hawai'i, the practice of counting by fours meant the terms equivalent to *one*, *ten*, *hundred*, *thousand*, and higher were *four*, *forty*, *four hundred*, *four thousand*, and so on,[12] suggesting the number system was quadragesimal (base-40), rather than decimal (base-10). Probably the greatest confusion associated with the practice of counting multiple items as the unit of counting was the story it generated, which claimed that New Zealand Māori counted by elevens (Chap. 31).

The counting method created an "ephemeral abacus." The abacus was *ephemeral* because it existed only as long as counting was being conducted, and it was an *abacus* because the different piles created by sorting had literal place value. However, objects like coconuts don't really lend themselves to the quick and easy movements needed for calculating. Thus, calculations were fairly limited: adding similar units (*two teau* and *two teau* would make *four teau*) or conjoining dissimilar units (*two teau* and *two afe* would make *two afe* and *two teau*, as ordered from largest units to smallest).[13] Nonetheless, counting as practiced reached into the thousands, and in most of Polynesia, counting numbers extended into the millions.

Another form of ephemeral abacus was created when different people kept track of the different units. One man would count things like the baskets of food brought to a feast, enumerating them from *one* to *ten*. Whenever he reached *ten*, a second man counted this to keep track of the tens; whenever the second man reached *ten*, a third man counted this to keep track of the hundreds, and so on.[14] This abacus was again ephemeral because it existed only as long as counting did, and it was again

[10] Von Chamisso (1821), pp. 441–442.

[11] Von Chamisso (1825), p. 7.

[12] Campbell (1816).

[13] Overmann (2021).

[14] Collocott (1925, 1927).

limited in terms of its potential for calculating. This kind of counting was also observed in Africa, as in the Sessuto numbers.

Such ephemeral forms are important because they resolve another longstanding mystery: how exponential structure (i.e., productive grouping by tens, hundreds, thousands, etc.) is created in societies that don't have an abacus. Historically, researchers have assumed that because some societies lack abacuses but have numbers with exponential structure, then numbers must be created by the unassisted brain, and these ideas are perhaps later externalized as material devices of some sort. In other words, first the ideas, then the material forms, and sometimes the second step just doesn't happen. But ephemeral forms show us two things: first, there are forms of abacuses, only they're temporary structures, not permanent devices. Second, a reversal of cause-and-effect is much more likely: that is, when we interact with material forms to rearrange their quantities, we get ideas about numbers. In other words, first the material forms, and then the ideas.

A binary form of counting-by-sorting anticipated Western binary computation by centuries.[15] In Mangareva, the large central island in the Gambier Islands in the eastern part of Polynesia, counting-by-sorting began in the same way as described before, but after the first round, the counters set aside were counted by eights, and then by fours, twos, and finally singles. Changing the basis of counting to *eight* helped to simplify the subsequent counting because it leveraged the binary relations between *one*, *two*, *four*, and *eight* (see Fig. 2.2). The presence of binary steps in a number system as highly elaborated as the one found throughout Polynesia is unusual. Possibly the practice emerged because in Mangareva, things were not just counted by ones, twos, and fours, but also by eights. One such object was the octopus, suggesting the practice was related to the animal's eight arms.[16] Whatever its antecedent, the practice extended the available binary relations, creating a four-step sequence.

As binary counting is found nowhere else in Polynesia, it most likely developed after Mangareva was settled between the eleventh and fourteenth centuries CE.[17] In the history of Western computing, binary numbers were invented in the late seventeenth century CE by the German mathematician Gottfried Leibniz, who wanted a notation that was short, unambiguous, easy to use, and revealing of numerical properties.[18] Base-2 helped realize modern computing because it reduced numbers to just two digits, 0 and 1, that could be represented with physical states: off and on. While the Mangarevan islanders didn't invent a computer, their binary counting nonetheless anticipated Leibniz's base-2 by as many as six centuries.[19]

---

[15] Bender and Beller (2014).

[16] Hiroa (1938); Janeau (1908).

[17] Rieth and Cochrane (2018).

[18] Couturat (1901).

[19] Bender and Beller (2014).

References

Andrews, L. (1854). *Grammar of the Hawaiian language*. Mission Press.

Aymonier, É. (1883). Quelques notions sur les inscriptions en vieux khmêr [Some notions on inscriptions in Old Khmer]. *Journal Asiatique, 8*(1), 441–505.

Bender, A., & Beller, S. (2014). Mangarevan invention of binary steps for easier calculation. *Proceedings of the National Academy of Sciences of the United States of America, 111*(4), 1322–1327.

Campbell, A. (1816). *A voyage round the world, from 1806 to 1812; in which Japan, Kamschatka, the Aleutian Islands, and the Sandwich Islands were visited.* Broderick and Ritter.

Chambers, G. K. (2013). Genetics and the origins of the Polynesians. In *eLS Wiley online library, evolution & diversity of life*. John Wiley & Sons.

Chambers, G. K., & Edinur, H. A. (2020). Reconstruction of the Austronesian diaspora in the era of genomics. *Human Biology, 92*(4), 247–263.

Clark, E. W. (1839). Hawaiian method of computation. *The Hawaiian Spectator, 2*(1), 91–94.

Cœdès, G. (1924). L'Épigraphie du Temple de Phimai [The epigraphy of the Temple of Phimai]. *Bulletin de l'École Française d'Extrême-Orient, 24*, 345–358.

Collocott, E. E. V. (1925). Supplementary Tongan vocabulary; also notes on measuring and counting, proverbial expressions and phases of the moon. *Journal of the Polynesian Society, 34*(3), 193–213.

Collocott, E. E. V. (1927). Kava ceremonial in Tonga. *Journal of the Polynesian Society, 36*(1), 21–47.

Cook, J. (1777). *A voyage towards the South Pole and round the world, performed in His Majesty's ships, the* Resolution *and* Adventure, *in the years 1772, 1773, 1774 and 1775. Vol. 2, Book III: From Ulietea to New Zealand*. W. Strahan and T. Cadell.

Couturat, L. (1901). *La logique de Leibniz: D'après des documents inédits [Leibniz's logic: Based on unpublished documents]*. Librairie Félix Alcan.

Guérin, V. (2017). The Oceanic subgroup of the Austronesian language family. In A. Y. Aikhenvald & R. M. W. Dixon (Eds.), *The Cambridge handbook of linguistic typology* (pp. 911–941). Cambridge University Press.

Hiroa, T. R. (1938). *Ethnology of Mangareva*. Bernice P. Bishop Museum.

Hurles, M. E., Matisoo-Smith, E., Gray, R. D., & Penny, D. (2003). Untangling Oceanic settlement: The edge of the knowable. *Trends in Ecology & Evolution, 18*(10), 531–540.

Janeau, V.-F. (1908). *Essai de grammaire de la langue des Iles Gambier ou Mangaréva par les Missionnaires catholiques de cet Archipel, membres de la Congrégation des Sacrés-Cœurs de Picpus [Essay on the grammar of the language of the Gambier or Mangareva Islands by the Catholic Missionaries of this Archipelago, members of the Congregation of the Sacred Hearts of Picpus]*. Imprimerie Zech et Fils.

Jenner, P. N. (1976). Les noms de nombre en Khmer [The names of the numbers in Khmer]. *Linguistics, 174*, 39–60.

Kirch, P. V. (1984). *The evolution of the Polynesian chiefdoms*. Cambridge University Press.

Leslie, J. (1820). *The philosophy of arithmetic; exhibiting a progressive view of the theory and practice of calculation, with an enlarged table of the products of numbers as far as one thousand* (2nd ed.). William and Charles Tait.

Lipson, M., Skoglund, P., Spriggs, M., et al. (2018). Population turnover in remote Oceania shortly after initial settlement. *Current Biology, 28*(7), 1157–1165.

Overmann, K. A. (2020). The curious idea that Māori once counted by elevens, and the insights it still holds for cross-cultural numerical research. *Journal of the Polynesian Society, 129*(1), 59–84.

Overmann, K. A. (2021). A new look at old numbers, and what it reveals about numeration. *Journal of Near Eastern Studies, 80*(2), 291–321.

Rieth, T., & Cochrane, E. E. (2018). The chronology of colonization in remote Oceania. In E. E. Cochrane & T. L. Hunt (Eds.), *The Oxford handbook of prehistoric Oceania* (pp. 133–161). Oxford University Press.

Roggeveen, J. (1721). Extract from the official log of Mynheer Jacob Roggeveen, in the ships Den Arend, Thienhoven and de Afrikaansche Galey, in 1721–2, in so far as it relates to the discovery of Easter Island. In B. G. Corney (Ed. & Trans.), *The voyage of Captain Don Filipe Gonzales to Easter Island, 1770–1* (pp. 1–26). Hakluyt Society.

Suggs, R. C. (1961). *Archaeology of Nuku Hiva, Marquesas Islands, French Polynesia*. American Museum of Natural History.

Von Chamisso, A. (1821). Corrections and remarks. In O. Von Kotzebue (Ed.), *A voyage of discovery, into the South Sea and Beering's Straits, for the purpose of exploring a north-east passage, undertaken in the years 1815–1818, at the expense of his Highness the Chancellor of the Empire, Count Romanzoff, in the ship* Rurick (Vol. III, pp. 439–442). Longman, Hurst, Rees, Orme, and Brown.

Von Chamisso, A. (1825). Du Grand Océan, de ses îles et de ses côtes [Of the Great Ocean, its islands and its coasts]. In L.-M. Bajot (Ed.), *Annales maritimes et coloniales, année 1825 [Maritime and colonial annals, year 1825]* (IIe partie, Tome 2, pp. 1–41). L'imprimerie Royale.

Von Chamisso, A. (1837). *Über die Hawaiische Sprache [About the Hawaiian language]*. Weidmannischen Buchhandlung.

Wynn, T., Overmann, K. A., & Malafouris, L. (2021). 4E cognition in the Lower Palaeolithic: An introduction. *Adaptive Behavior, 29*(2), 99–106.

Chapter 31
Misunderstandings, Mistakes, and Outright Frauds

Abstract Traditional Polynesian counting practices and numerical vocabularies have been significantly misunderstood or misrepresented by Western observers. In the late eighteenth century, Spanish seafarers collected a list of "number-words" from Rapa Nui islanders by showing them written Western numerals. Unsurprisingly, the results bear no resemblance to the Polynesian words for numbers collected in Rapa Nui by Captain Cook (and everyone else who followed him) only a few years later. A few years after the Rapa Nui incident, a Frenchman collected and published a list of Tongan "number-words" that turned out to be a crude practical joke on him. In the early nineteenth century, two French naturalists who had visited New Zealand spread the story that Māori counted by elevens, something no natural number system has ever been known to do. The hoax appears to have started as a printer's error, but it intensified when the two men wrote to colleagues with additional (fabricated) details. In the late twentieth century, an amateur linguist published a collection of Pacific numerical vocabularies; about half were legitimate vocabularies, while the remainder were clumsy forgeries.

Keywords Rapa Nui · Labillardière · Undecimal counting · René Lesson · Peter Lanyon-Orgill

One of the most fascinating aspects of number systems in Polynesia is their potential for being misunderstood. Visitors to the region have sometimes misinterpreted what they were seeing; this is reasonably explained—especially in the initial centuries of contact—as newcomers being unfamiliar with language and customs, and number systems that are organized differently being difficult to understand. Other times, visitors were simply mistaken, sometimes hilariously so. And on occasion, someone perpetrated a fraud.

The Spanish collected a mysterious list of "number-words" from Rapa Nui. In 1770, Spanish seafarers made the second Western contact with Rapa Nui (also

K. A. Overmann, *Cultural Number Systems*, Interdisciplinary Contributions to Archaeology, https://doi.org/10.1007/978-3-031-83383-0_31

known as Easter Island), a Dutch visit some 50 years earlier being the first.[1] During the visit, the Spanish collected vocabulary from the islanders, eliciting words through pantomime, either pointing to things or enacting them. In this way, they obtained words like hair ("Coojo"), man ("Cotajata"), sky ("Gerani"), to sleep ("Geuru"), and to look at ("Arà").[2] When it came to the words for numbers, however, the Spanish apparently showed the islanders the written numerals 1 to 10. While these symbols clearly meant numbers to the Spanish, the islanders had never seen them before, though of course they had counting devices like tallies and most likely knew the method of counting-by-sorting found throughout Polynesia (Chap. 30). The islanders also had a form of writing, an as-yet undeciphered script known as Rongorongo[3] that—unusually—seems not to include any symbols for numbers.[4] While the Western numerals did indeed elicit vocabulary from the islanders, the results bore no resemblance whatsoever to the typical Polynesian words for numbers that Captain Cook would collect in Rapa Nui only a few years later.[5] Researchers have been trying ever since to make numbers out of the words collected during the Spanish visit, without much success.[6]

An early hoax concerned the Tongan words for numbers. Jacques de Labillardière was a member of the 1791–1792 expedition that searched for the *La Pérouse* expedition, two French ships that had mysteriously vanished somewhere in the Pacific in 1788. Despite the expedition's serious purpose and the personal interest of King Louis XVI, whose government had ordered the rescue mission,[7] Labillardière nonetheless found time to quiz the Tongans on their words for numbers; the list he collected was published in 1799 and translated into English in 1800.[8] The list contained the names of very high numbers—up to *ten* to the fifteenth power, a quadrillion. Instead of words for numbers, however, some of the words were literally "nonsense," while others were obscene, so indelicate they were not included in the descriptive paragraph, but were instead kept sequestered in the vocabulary listed in the volume's final pages, and even there they were decorously covered in Latin.[9] The word for 1,000,000 had "no meaning that we can discover"; 10,000,000 was "the praeputium," the foreskin of the penis; 100,000,000 was the word for "*nonsense*" in the Tongan language; 1,000,000,000 was "the name of a game"; 10,000,000,000 and 100,000,000,000 were combined as a phrase meaning "Pudendum muliebre," the external part of the female genitalia (literally, "a woman's thing to be ashamed of"); 1,000,000,000,000 meant "the male organs of

[1] Roggeveen (1721).

[2] González de Haedo (1770, p. 110).

[3] Fedorova (1993); Fischer (1997).

[4] Davletshin (2012).

[5] Cook (1777).

[6] Fedorova (1969); Mellén Blanco (1993); Métraux (1936); Ross (1936).

[7] Hitchcock (2017).

[8] De Labillardière (1799, 1800).

[9] Martin (1818).

generation (a vulgarism)"; 10,000,000,000,000 was "the glans penis"; and 100,000,000,000,000 and 1,000,000,000,000,000 were instructions "to eat up the things which they [the Tongans] have just been naming to him [Labillardière]."[10]

A contemporary reader, presumably versed in Latin and curious enough to look up the vocabulary at the back of the volume, encountered Labillardière's book, which had quickly found its way to the library of Sir Joseph Banks, a gentleman-naturalist who had accompanied Captain James Cook on his first Pacific voyage (1768–1771) and would become the president of the Royal Society for over four decades and an advisor to King George III.[11] The reader's reaction to the list of "numbers" is worth repeating:

> A young missionary, who was reading a book of travels in Sir Joseph Banks's library, was observed every now and then to burst out into a violent fit of laughter; and on the cause of this being asked, it was found that he was reading over a vocabulary in which the natives had cheated the scientific compiler, by giving such answers to his inquiries, that, had any future voyager attempted to use the work of his predecessor, no very good opinion would have been entertained of his morals, and he would have been far distant from the attainment of any object for which he might think he was asking.
> Craik (1830, p. 417)

The incident was published by the Scottish author George Lillie Craik, who observed on the same page that the South Sea Islanders were known "to amuse themselves with that sort of wit which [Jonathan] Swift calls 'selling a bargain'."[12] It's delightful to find the local inhabitants were having the occasional bit of fun with the naïve foreigners who visited their island.

Probably the biggest hoax was the story that Māori counted by elevens. Confused by counting with pairs, the German botanist Adelbert von Chamisso described the New Zealand Māori number system as *vigesimal* (or base-20) in an 1821 publication,[13] when it is actually a decimal system (base-10). In 1825, another gentleman-scientist of the nineteenth century, René Lesson, translated this work by von Chamisso's into French.[14] Noting the erroneous description, Lesson added a footnote to the effect that the New Zealand number system was "undecimal," or based on the number *eleven*.[15] This was possibly a printer's error that conjoined the words "un decimal," which would have correctly identified Māori numbers as decimal.[16] Lesson certainly knew Pacific numbers were decimal, as he had visited the Pacific from 1822 to 1825 as a naturalist onboard the *Coquille*, a one-time horse barge of the French navy pressed into new service as an exploring vessel, and in his

---

[10] De Labillardière (1799), p. 50; Martin (1818), pp. 370–371.

[11] Dryander (1800); Moore (2009).

[12] Craik (1830), p. 417.

[13] Von Chamisso (1821).

[14] Von Chamisso (1825).

[15] Lesson, in Von Chamisso (1825), p. 27, footnote 2.

[16] Overmann (2020).

capacity as naturalist, he had collected numerical vocabularies and ultimately published about a dozen of them on his return to France.[17]

After the erroneous footnote was published, Lesson and his friend Jules de Blosseville, who had been the *Coquille*'s navigator on the same voyage, began writing to other gentlemen-scientists claiming that Māori did indeed count by elevens. They described an elevens-based counting method and the Māori informants who had allegedly demonstrated it, and they furnished a purported vocabulary with terms for *eleven-squared* and *eleven-cubed* (the base-11 counterparts to *hundred* and *thousand*).[18] A few of these letters survived, published second- or thirdhand by their recipients; some were attributed to Lesson or de Blosseville by name,[19] while others were anonymous but thought to have been written by Lesson.[20] Interestingly, neither Lesson nor de Blosseville published the story themselves, though both men published multiple accounts of their travels with the *Coquille*, many of which featured New Zealand.[21]

Because counting by elevens is so unusual, the idea that Māori did so would be repeated for decades,[22] with varying degrees of credulity. Over time, the story began losing its details; by the middle of the twentieth century, the "undecimal" vocabulary had been lost, René Lesson's role as chief instigator was unknown, and the story itself had nearly been forgotten—until new research revealed the possible mistake on the part of the printer and subsequent mischief by Lesson and de Blosseville.[23]

A third hoax was much more recent, and it spanned the whole Pacific region. In 1979, Peter Lanyon-Orgill (1924–2002), an enthusiastic but uncredentialed linguist who claimed (but apparently never earned) a doctorate, published a collection of numerical vocabularies from around the Pacific region. Half were legitimate vocabularies he had gathered from obscure sources, by means and methods unknown.[24] The other half he claimed were taken from "a small volume [found] among the papers of [his] distinguished ancestor, William Lanyon,"[25] who reportedly served on the three voyages made by Captain Cook in the years 1768–1771, 1772–1775, and 1776–1780. In actuality, there is no record of Lanyon-Orgill being related to the William Lanyon who sailed with Cook, no record of William Lanyon collecting or publishing any numerical vocabularies,[26] and historical records showing Cook never visited or collected words from some of the places the vocabularies were allegedly

[17] E.g, Lesson (1825a, b); Lesson and Garnot (1826).

[18] Balbi (1826); Von Humboldt (1839); Von Zach (1826).

[19] Craik (1830); Von Zach (1826).

[20] Von Humboldt (1839); also see Pott (1847).

[21] Overmann (2020).

[22] E.g, Conant (1896); Williams (1844, 1852).

[23] Overmann (2020).

[24] Clark (2011); Geraghty (1983).

[25] Lanyon-Orgill (1979), p. ix.

[26] Clark (2011).

from.[27] When asked for proof in the form of the original manuscript, Lanyon-Orgill claimed it was missing—seized, miscatalogued, and then sold and resold as part of a bankruptcy proceeding.[28]

Lanyon-Orgill's 1979 volume was revealed as a fraud within just a few years of its publication. In 1983, a critical review in *The Journal of the Polynesian Society* exposed factual errors, historical anachronisms, and linguistic mistakes in half of the volume's vocabularies.[29] The review led to the realization that "fraudulent elements"—"imaginary friends," untraceable sources, unverifiable credentials and fieldwork[30]—had characterized Lanyon-Orgill's work throughout his entire career as a linguist. Later scholars have occasionally fallen victim, using the falsified vocabularies in the belief they were legitimate historical data.[31]

References

Balbi, A. (1826). Observations sur la classification des langues océaniennes [Observations on the classification of Oceanic languages]. In *Atlas ethnographique du globe, ou classification des peuples anciens et modernes d'après leur langue. Vol. 1: Discours préliminaire et introduction [Ethnographic atlas of the globe, or Classification of ancient and modern peoples according to their language. Vol. 1: Preliminary discourse and introduction]* (pp. 230–278). Paul Renouard.

Clark, R. (2011). On the margins of Pacific linguistics: P. A. Lanyon-Orgill. *Language and History, 54*(2), 164–177.

Conant, L. L. (1896). *The number concept: Its origin and development*. Macmillan.

Cook, J. (1777). *A voyage towards the South Pole and round the world, performed in His Majesty's ships, the* Resolution *and* Adventure, *in the years 1772, 1773, 1774 and 1775. Vol. 2, Book III: From Ulietea to New Zealand*. W. Strahan and T. Cadell.

Craik, G. L. (1830). *The New Zealanders*. Wells and Lilly.

Davletshin, A. (2012). Numerals and phonetic complements in the "Kohau Rongorongo" script of Easter Island. *Journal of the Polynesian Society, 121*(3), 243–274.

De Labillardière, J. J. H. (1799). Appendix: Vocabulaire de la langue des Iles des Amis [Appendix: Vocabulary of the language of the Friendly Islands]. In *Relation du voyage a la recherche de* La Pérouse, *fait par ordre de l'Assemblee Constituante, pendant les années 1791, 1792, et pendant la 1ère et la 2de année de la République Françoise [Report of the voyage in search of* La Pérouse, *made by order of the Constituent Assembly, during the years 1791, 1792, and during the 1st and 2nd years of the French Republic]* (Vol. II, pp. 41–50). H. J. Jansen.

De Labillardière, J. J. H. (1800). Vocabulary of the language of the Friendly Islands. In *Voyage in search of* La Pérouse. *Translated from the French* (Vol. II, pp. 47–58). John Stockdale.

Dryander, J. (1800). *Catalogus bibliothecæ historico-naturalis Josephi Banks. Tomus V: Supplementum et index auctorum [Catalog of the natural history library of Joseph Banks. Vol. V: Supplement and list of authors]*. Bulmer et Soc.

Fedorova, I. K. (1969). О десяти числительных рапануйского языка [About ten numerals of Rapanui language]. *Sovietskaia Etnografia, 1969*(1), 144–146.

[27] Clark (2011); Geraghty (1983).

[28] Lanyon-Orgill (1979).

[29] Geraghty (1983).

[30] Clark (2011), pp. 164, 173.

[31] Clark (2011).

Fedorova, I. K. (1993). The Rapanui language as a source of ethnohistorical information. *Rongorongo Studies, 3*(2), 52–60.

Fischer, S. R. (1997). *Rongorongo: The Easter Island script*. Clarendon Press.

Geraghty, P. (1983). Review, Lanyon-Orgill, Peter A.: *Captain Cook's South Sea island vocabularies*. London, Peter Lanyon-Orgill, 1979. *Journal of the Polynesian Society, 92*(4), 554–559.

González de Haedo, D. F. (1770). *The voyage of Captain Don Felipe Gonzalez in the ship of the line* San Lorenzo, *with the frigate* Santa Rosalia *in company, to Easter Island in 1770–1771* (B. G. Corney, Ed. & Trans.). Hakluyt Society.

Hitchcock, G. (2017). The final fate of the *La Pérouse* Expedition? The 1818 account of Shaik Jumaul, a Lascar castaway in Torres Strait. *Journal of Pacific History, 52*(2), 217–235.

Lanyon-Orgill, P. A. (1979). *Captain Cook's South Sea island vocabularies*. Author.

Lesson, R.-P. (1825a). *Notice sur l'île de Oualan ou Strong [Notice on the island of Oualan or Strong]*. Imprimerie de Gœtschy.

Lesson, R.-P. (1825b). Notice sur l'île de Rotouma, située dans le Grand-Océan austral [Notice on the island of Rotouma, located in the Great Southern Ocean]. In J.-B. B. Eyriès & C. Malte-Brun (Eds.), *Nouvelles annales des voyages, de la géographie et de l'histoire [New annals of travel, geography and history]* (Vol. XXVII, pp. 1–47). Librairie de Gide Fils.

Lesson, R.-P., & Garnot, P. (1826). *Voyage autour du monde, exécuté par ordre du Roi, sur la Corvette de Sa Majesté,* La Coquille, *pendant les années 1822–1825: Zoologie [Voyage round the world, executed by order of the King, on His Majesty's Corvette, the* Coquille, *during the years 1822–1825: Zoology]* (L. I. Duperrey, Ed., Vol. 1). Arthus Bertrand.

Martin, J. (1818). *An account of the natives of the Tonga Islands, in the South Pacific Ocean. With an original grammar and vocabulary of their language. Compiled and arranged from the extensive communications of Mr. William Mariner, several years resident in those islands* (Vol. II). John Murray.

Mellén Blanco, F. (1993). Interpretación Rapa Nui de los numerales escritos por los españoles en 1770 [Rapa Nui interpretation of the numerals written by the Spanish in 1770] (S. Hawson, Trans.). *Revista Española Del Pacífico, 3*, 61–66.

Métraux, A. (1936). Numerals from Easter Island. *Man, 36*(253–254), 190–191.

Moore, H. (2009). *Joseph Banks*. Heinemann Library.

Overmann, K. A. (2020). The curious idea that Māori once counted by elevens, and the insights it still holds for cross-cultural numerical research. *Journal of the Polynesian Society, 129*(1), 59–84.

Pott, A. F. (1847). *Die quinare und vigesimale Zählmethode bei Völkern aller Welttheile: Nebst ausführlicheren Bemerkungen über die Zahlwörter Indogermanischen Stammes und einem Anhange über Fingernamen [The quinary and vigesimal counting method among peoples of all parts of the world: With detailed remarks on the number words of Indo-European tribes and an appendix on finger names]*. C. A. Schwetschke und Sohn.

Roggeveen, J. (1721). Extract from the official log of Mynheer Jacob Roggeveen, in the ships *Den Arend, Thienhoven* and *de Afrikaansche Galey*, in 1721–2, in so far as it relates to the discovery of Easter Island. In B. G. Corney (Ed. & Trans.), *The voyage of Captain Don Filipe Gonzales to Easter Island, 1770–1* (pp. 1–26). Hakluyt Society.

Ross, A. S. C. (1936). Preliminary notice of some late eighteenth century numerals from Easter Island. *Man, 36*(120), 94–95.

Von Chamisso, A. (1821). Corrections and remarks. In O. Von Kotzebue (Ed.), *A voyage of discovery, into the South Sea and Beering's Straits, for the purpose of exploring a north-east passage, undertaken in the years 1815–1818, at the expense of his Highness the Chancellor of the Empire, Count Romanzoff, in the ship* Rurick (Vol. III, pp. 439–442). Longman, Hurst, Rees, Orme, and Brown.

Von Chamisso, A. (1825). Du Grand Océan, de ses îles et de ses côtes [Of the Great Ocean, its islands and its coasts]. In L.-M. Bajot (Ed.), *Annales maritimes et coloniales, année 1825 [Maritime and colonial annals, year 1825]* (IIe partie, Tome 2, pp. 1–41). L'imprimerie Royale.

Von Humboldt, W. (1839). *Über die Kawi-Sprache aus der Insel Java, nebst einer Einleitung über die Verschiedenheit des menschlichen Sprachbaues und ihren Einsluss aus die geistige Entwickelung des Menschengeschlechts. Band III. Südsee-Sprachen, als östlicher Zweig des Malayischen [On the Kawi language from the island of Java, together with an introduction on the diversity of human language structure and its influence on the intellectual development of the human race. Vol. III: South Sea languages, as an eastern branch of Malay]*. F. Dümmler.

Von Zach, F. X. (1826). Correspondance astronomique, géographique, hydrographique et statistique [Astronomical, geographical, hydrographic and statistical correspondence]. In E. Saigey (Ed.), *Bulletin des sciences mathématiques, astronomiques, physiques et chimiques [Bulletin of mathematical, astronomical, physical and chemical sciences]* (1ère section, Vol. 5, pp. 120–123). Bureau du Bulletin.

Williams, W. (1844). *A dictionary of the New-Zealand language, and a concise grammar; to which are added a selection of colloquial sentences*. The Press of the Church Mission Society.

Williams, W. (1852). *A dictionary of the New Zealand language, and a concise grammar; to which are added a selection of colloquial sentences* (2nd ed.). Williams and Norgate.

Part VI
Number Systems in the Americas

Chapter 32
The Arctic: Inuit Numbers

Abstract Eskaleut number systems are quinary-vigesimal, counting by groups of *five* and *twenty*. Such numbers reflect counting with the hands and feet, as illustrated by the names given to the numbers in the Greenland Inuit number system. Inuit numbers accumulate, with a few subtractive (back-counting) formations. These systems show that counting with the toes isn't limited to the tropics, as is widely believed, since it is also practiced in the coldest places on the planet. Counting beyond *twenty* recruited the hands and feet of other people or bundles of household items. Bundles also condensed counting in a way that mitigated the need to develop terms for higher numbers. Inuit numbers have been undergoing a digital revitalization over the past several decades. Indigenous number systems are often displaced by Western numerals through processes of contact, education, and political incorporation. In the mid-90s, to resist this trend, Inuit students and their teacher invented a set of written numerals, named "Kaktovik numerals" after their Alaskan village of origin. Kaktovik numerals are a quinary-vigesimal system with place value and *zero*.

Keywords Quinary · Vigesimal · Back-counting · Kaktovik numerals · Digital revitalization

The Arctic is the snow-covered region at the planet's northern pole, as delimited by the southernmost extent of the visibility of the midnight sun at the summer solstice and its invisibility at the winter solstice. The region encompasses most of the Arctic Ocean and the northernmost parts of Alaska, Canada, Greenland, Scandinavia, and Siberia. Average temperatures there range from below −40 °F in winter to above 50 °F in summer.[1] The Arctic is home to many peoples, including the Aleut (or Unangan), Inuit, and Yupik. Their languages make up the Eskaleut language family but are mutually unintelligible except to near-neighbors; Aleut is spoken in Alaska

[1] *Arctic Weather and Climate* (2024).

K. A. Overmann, *Cultural Number Systems*, Interdisciplinary Contributions to Archaeology, https://doi.org/10.1007/978-3-031-83383-0_32

and the Aleutian Islands, Inuit in northern Alaska, Arctic Canada, and Greenland, and Yupik in Alaska and Siberia.[2]

Eskaleut number systems are quinary-vigesimal, counting by groups of five *and* twenty. As might be surmised, such numbers reflect counting with the hands and feet, as attested by the names given to the numbers. An example is the Greenland Inuit number system (Table 32.1):

The constructions "at the edge" mean that counting shifts to the next appendage: over to the next hand or foot, and down from hands to feet.[3] However, not all Eskaleut languages continue counting up when shifting to the nearest hand or foot. In Iñupiaq, the North Alaskan Inuit language, the numbers *nine*, *fourteen*, and *nineteen* subtract *one* from the next higher multiple of *five*, a type of back-counting. Thus, *nine* is *quliŋŋuʁutail*^{j}*aq*, or "*ten* minus *one*"; *fourteen* is *akimaʁutail*^{j}*aq*, or "*fifteen* minus *one*"; and *nineteen* is *iñuiññaʁutail*^{j}*aq*, or "*twenty* minus *one*."[4] These terms can have different translations: One source gives *qulingiluaqtuq* (a variant of the word for *nine*) as "it isn't ten."[5]

Table 32.1 Greenland Inuit numbers (compiled from Dorais, 2010; Fortescue, 2008; and MacLean, 2014)

Nr.	Greenland Inuit	Meaning
1	*ataaseq*	"that which persists/remains"
2	*marluk*	"[the pair] following"
3	*piŋasut*	"the ones [fingers] going up" (i.e., to the middle finger)
4	*sisamat*	"those [fingers] up to the hard one" (i.e., the index finger)
5	*tallimat*	"those [fingers] making up the [whole] arm"
6	*arfinillit*	"[it goes across (to)] the edge of the [other] hand"
7	*arfineq marluk*	"[having reached] the edge [plus] [the pair] following"
8	*arfineq piŋasut*	"[having reached] the edge [plus] three [fingers]"
9	*arfineq sisamat*	"[having reached] the edge [plus] four [fingers]"
10	*qulit*	"[all] the ones [fingers] in the upper part [of the body]"
11	*aqqanillit*	"it goes down [to the feet]"
12	*aqqaneq marluk*	"descent [plus] [the pair] following"
13	*aqqaniq piŋasut*	"descent [plus] three [toes]"
14	*aqqaneq sisamat*	"descent [plus] four [toes]"
15	*aqqaneq tallimat*	"descent [plus] five [toes]"
16	*arfersanillit*	"[it goes across (to)] the edge of the [other] foot"
17	*arfersaneq marluk*	"[having reached] the edge [plus] [the pair] following"
18	*arfersaneq piŋasut*	"[having reached] the edge [plus] three [toes]"
19	*arfersaneq sisamat*	"[having reached] the edge [plus] four [toes]"
20	*inuk naallugu*	"completing the person"

[2] Bergsland (2022).

[3] Denny (1986), p. 134.

[4] Kaplan (2007).

[5] McKitrick (2021), p. 46.

Using the hands for counting is universal among human societies, and involving the feet is not rare. What may be surprising, however, is that involving the feet isn't limited to the tropics, as is widely believed, since the Arctic number systems show that it is also practiced in the coldest places on the planet. Nonetheless, given the Arctic cold and potential for frostbite, traditional counting wouldn't often involve exposing the extremities to the open air—not just the toes, but also the fingers. Instead, counting would tend to be performed indoors, which would also tend to make the practice highly social. Such communality means that counting beyond *twenty*, the number of digits on a single human body, was also likely to involve the digits of other people.

Counting beyond twenty *recruited the hands and feet of other people*. For the Greenland Inuit, *twenty-one* was *inûp áipagssâne atausek*, meaning "*one* [finger] on the second man"; *thirty-eight* was *inûp áipagssâne arfinek pingasut*, "three toes on the second man's second foot"; and *forty* was *inûp áipagssâ nâvdlugo*, "the whole of the second man."[6] Recruiting the digits of other people is a common strategy for extending the upper limit of a counting sequence based on counting with the hands (decimal) or hands and feet (vigesimal). A similar example is found in Central Africa, where counting beyond twenty for the Zande (or Azande) meant "another person is requisitioned, and the reckoner proceeds to count his fingers and toes."[7]

Counting beyond twenty *could also involve bundles of household items*. This too is a common strategy. As noted in an Iñupiaq dictionary, "The wife and mother in any Eskimo household knew how many bundles were required for a particular purpose, such as sewing a parka."[8] In the village of Nuchek, Alaska, the word *qangaanaq*—literally, "squirrel"—was used to express *one hundred*: "It is said that in past days a bundle of 45 squirrels was needed to make a parka in some areas, and a bundle of 100 squirrels was probably needed in areas like Nuchek as the squirrels in this village were smaller than in other villages."[9] The same strategy is found in West Africa, where big numbers like *four hundred* (or *twenty* squared, the counterpart of *hundred* in a decimal system) were named after bundles of cowrie shells; in Mesoamerica, where the Maya named some of their big numbers after bundles of cacao beans; and in South America, where Mochica, a now-extinct language spoken in northern Peru at the time of the Spanish conquest, certain types of objects were counted by pairs, tens, or hundreds.[10]

As a general rule, lengthy names for numbers that are "etymologically transparent" (or easily analyzable) as digit-counting, like those in Table 32.1, indicate that a number system is likely recent in origin, used infrequently, or both. This is especially true for number systems whose upper limit is low (e.g., *twenty*, the limit of

[6] Nansen (1893), pp. 194–195.

[7] Gore (1926), p. 42.

[8] MacLean (2014), p. 832.

[9] Koo (1980), p. 286.

[10] Bellamy (2018).

"restricted" counting systems[11]). However, this rule isn't easily applied to Inuit number systems, for two reasons. First, the number names in Table 32.1, while multisyllabic,[12] are well shortened, as indicated by brackets around meaning in the glosses that is understood rather than explicit pronounced; such shortening is indicative of both longevity and frequent use. Second, counting bundles of multiple items, rather than each item individually, condenses counting in a way that eliminates the need for higher numbers. For example, an Inuit family might store 400 individual fish for the upcoming winter, but this number isn't nearly as meaningful as the total mass of fish, which is more accurately assessed by counting the number of racks used for drying and storing the fish.[13] Counting collective units like racks for fish means that counting sequences are under far less pressure to extend their upper limit.

Inuit numbers are undergoing a digital revitalization. Indigenous number systems are often displaced by Western numerals through processes of contact, education, and political incorporation. In the mid-90s, to resist this trend, Inuit students and their teacher invented a set of written numerals, called "Kaktovik numerals" (Fig. 32.1) after their Alaskan village of origin.[14]

The Kaktovik numerals are a quinary-vigesimal system with place value and *zero*. The iconography for *zero* is based on the crossed arms, a gesture meaning

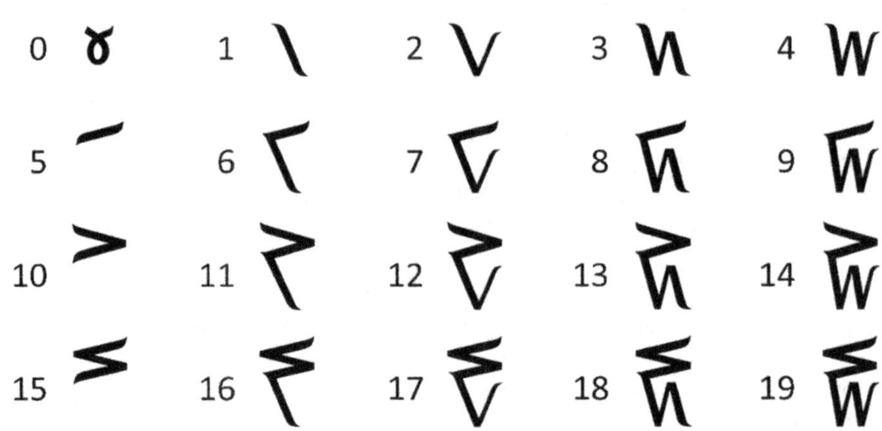

Fig. 32.1 Kaktovik numerals. Of note, the numerals 9, 14, and 19 follow the additive rule shown in Table 32.1 (the lower multiple of 5 + 4), rather than the subtractive rule used in the Iñupiaq language spoken by the inventors (the next higher multiple of 5 – 1). Adapted and labeled from MacLean (2014, Table 1, p. 832)

[11] Comrie (1989, 2013).

[12] "Notes on Inuktitut Pronunciation," (2024).

[13] MacLean (2014).

[14] Bartley (1997).

"nothing."[15] The iconography for the other numerals thoughtfully combines vertical strokes for the numbers 1 through 4 with horizontal strokes for the multiples of 5.[16] The number 20 is 𝋁𝋀, the symbol for *one* in the twenties' place and the symbol for *zero* in the units' place. The forms of the symbols visually aid the performance of arithmetic in a way that isn't possible with Western numerals, where arithmetic depends on memorized numerical relations. For example, subtracting 2 from 12 (𝋌 − 𝋂) simply removes the vertical elements for 2 (𝋂) from the symbol for 12 (𝋌), leaving the horizontal elements for 10 (𝋊). Similarly, adding 5 to 12 (𝋌 + 𝋅) means adding the third horizontal element that changes 10 (𝋊) to 15 (𝋏), yielding 17 (𝋑). With small divisors like 2, 3, or 4, division becomes a straightforward matter of removing that number of horizontal and vertical elements from the dividend.

After inventing the numerals and a base-20 abacus, Inuit students worked intensively with both the Western decimal and Kaktovik vigesimal systems, improving their standardized math scores "from below the 20th percentile to 'significantly above' the national average" in just a single year.[17] Despite this demonstrated value, the Kaktovik numerals were subsequently relegated to Iñupiaq language classrooms for over a decade when their use was seen as outside of the requirements of the 2001 No Child Left Behind Act.[18] The situation began to change in 2022 when the numerals were included as Unicode characters; as of 2024, they have become accessible as fonts, online converters, and iPhone apps, greatly increasing the digital availability critical to computerized environments.[19] These developments promise to do more than the already-important job of furnishing Indigenous classrooms with culturally relevant materials: The human capacity for numbers and numeracy is much broader than the narrow confines of the Western tradition, and bringing Indigenous number systems into the digital age holds enormous potential to expand our appreciation and understanding of this capacity.

The Kaktovik numerals evoke an earlier invention. Around 1820, a Cherokee named Sequoyah invented a syllabary (a set of symbols for syllables) for writing his language.[20] Around 1830, not quite a decade before the Cherokees would be forcibly removed west of the Mississippi along the infamous "Trail of Tears,"[21] Sequoyah also created a set of numerals. While the syllabary was widely adopted and remain in use today, the numerals were not, and today are mostly known from two surviving letters.[22] The lack of success is attributed to their complexity, a property the simpler Kaktovik numerals fortunately do not share.

---

[15] Tillinghast-Raby (2023), p. 14.

[16] McKitrick (2021).

[17] Bartley (1997); Tillinghast-Raby (2023).

[18] Tillinghast-Raby (2023).

[19] Binggeli (2023); Tillinghast-Raby (2023).

[20] Chrisomalis (2010).

[21] Hausman (2011); Thornton (1984).

[22] Chrisomalis (2010, 2020).

References

Arctic weather and climate. (2024). *National Snow and Ice Data Center*. https://nsidc.org/learn/parts-cryosphere/arctic-weather-and-climate

Bartley, W. C. (1997). Making the old way count. *Sharing our pathways: A newsletter of the Alaska Rural Systemic Initiative, 2*(1), 12–13.

Bellamy, K. (2018). Re-evaluating the 'numeral classifier' system in Mochica. *LIAMES: Línguas Indígenas Americanas, 18*(1), 81–98.

Bergsland, K. (2022). Eskimo-Aleut languages. In *Encyclopaedia Britannica*. https://www.britannica.com/topic/Eskimo-Aleut-languages

Binggeli, M. (2023). *Digitization of Kaktovik numbers*. https://polarjournal.ch/en/2023/04/18/digitization-of-kaktovik-numbers/%0APolarjournal

Chrisomalis, S. (2010). *Numerical notation: A comparative history*. Cambridge University Press.

Chrisomalis, S. (2020). *Reckonings: Numerals, cognition, and history*. MIT Press.

Comrie, B. (1989). *Language universals and linguistic typology* (2nd ed.). University of Chicago Press.

Comrie, B. (2013). Numeral bases. In M. S. Dryer & M. Haspelmath (Eds.), *World atlas of language structures online*. Max Planck Institute for Evolutionary Anthropology. http://wals.info/chapter/131

Denny, J. P. (1986). Cultural ecology of mathematics: Ojibway and Inuit hunters. In M. P. Closs (Ed.), *Native American mathematics* (pp. 129–180). University of Texas Press.

Dorais, L.-J. (2010). *The language of the Inuit: Syntax, semantics, and society in the Arctic*. McGill-Queen's University Press.

Fortescue, M. (2008). Greenlandic Inuktitut, Greenland. In E. Chan (Ed.), *Numeral systems of the world's languages*. Max Planck Institute. https://lingweb.eva.mpg.de/channumerals/Greenlandic-Inuktitut.htm

Gore, E. C. (1926). *A Zande grammar*. Sheldon Press.

Hausman, B. M. (2011). *Riding the Trail of Tears*. University of Nebraska Press.

Kaplan, L. (2007). North Alaskan Iñupiaq, Alaska. In E. Chan (Ed.), *Numeral systems of the world's languages*. Max Planck Institute. https://lingweb.eva.mpg.de/channumerals/Inupiatun-North-Alaskan.htm

Koo, J. H. (1980). The number system of Aleut Eskimo. *Language Research, 16*(2), 281–288.

MacLean, E. A. (2014). *Iñupiatun uqaluit Taniktun sivuniŋit/Iñupiaq to English dictionary*. University of Alaska Press.

McKitrick, P. A. (2021). *A guide to teaching in Nunavut*. Unpublished master's thesis, Lakehead University.

Nansen, F. (1893). *Eskimo life* (William Archer, Trans.). Longmans, Green, and Co.

Notes on Inuktitut pronunciation. (2024). In *Inuit nipingit/Inuit sounds*. Inhabit Media. https://inhabitmedia.com/inuitnipingit/

Thornton, R. (1984). Cherokee population losses during the Trail of Tears: A new perspective and a new estimate. *Ethnohistory, 31*(4), 289–300.

Tillinghast-Raby, A. (2023). Arctic arithmetic. *Scientific American, 328*(6), 14–16.

Chapter 33
Northern California: Pomo Numbers

Abstract The Pomo became comparatively wealthy among the Californian tribes by making small cylindrical beads from clamshells. When strung, these beads served as a standard of currency throughout the region. The Pomo counted strung beads into the tens of thousands, as documented through the names given to the numbers and a few ethnographic observations. No trace of the ancient Pomo number system is found archaeologically, illustrating a key challenge in investigating prehistoric number systems. The practice of counting beads was related to, and presumably motivated by, the value the beads achieved from the amount of labor it took to produce them, along with their widespread use as standard currency. The practice of counting beads, along with the need to count them in significant quantities, meant the Pomo number system expanded to very high numbers. The names of the Pomo numbers were directly related to the way the beads were strung and then bundled for the purpose of counting them. The Yuki, one of the Pomo's near neighbors, had a number system based on the number *eight*, along with a unique decimal-octal counting sequence presumably used with beads.

Keywords Clamshell beads · Strung beads · Currency · Yuki · Octal

The Pomo are Americans native to northern California. Their historical territory, which they have inhabited for thousands of years,[1] encompasses the region today called Sonoma County, now known for its wine production. Pomo territory was centered on the Russian River Valley some 50–100 miles north of San Francisco, and it included the adjacent coastal and inland regions.[2] Among the Californian tribes, the Pomo became comparatively wealthy by making small cylindrical beads from clamshells (Fig. 33.1). When strung, these beads "served as a medium of

[1] Basgall (1982).

[2] Gorlinski (2019).

K. A. Overmann, *Cultural Number Systems*, Interdisciplinary Contributions to Archaeology, https://doi.org/10.1007/978-3-031-83383-0_33

Fig. 33.1 Pomo clamshell beads. Adapted from MoneyMuseum (2024) and used with permission from MoneyMuseum, Zurich

exchange, a standard of value, and a means of storing wealth" throughout north-central California.[3]

The Pomo counted beads into the tens of thousands. This practice was related to, and presumably motivated by, the value the beads achieved from the amount of labor it took to produce them, along with their widespread use as standard currency.[4] The beads were made from the shells of saltwater clams, known as *ká yah* or "water bone."[5] The shells were "first broken in the hand with a stone into roughly rounded disks, and later by means of a quartz implement having sharp edges, the angles [were] deftly chipped off"; the beads were then drilled, polished, and "sorted into strings according to [their] diameter."[6]

Importantly, the motivation to count these valuable items, coupled with the need to count them in significant quantities, meant the Pomo number system expanded to very high numbers—whose names were directly related to the way the beads were strung and then bundled for the purpose of counting them. It's worth emphasizing that beads were not counted one-by-one to high numbers. Rather, small quantities of beads were strung, the strings were then bundled, and bundles of increasing size were used to represent the higher numbers.

[3] McLendon and Lowy (1978), p. 311.

[4] Loeb (1926), p. 230, drawing on material from Kroeber (1925), pp. 256–257.

[5] Hudson (1897), p. 104.

[6] Hudson (1897), pp. 104–105

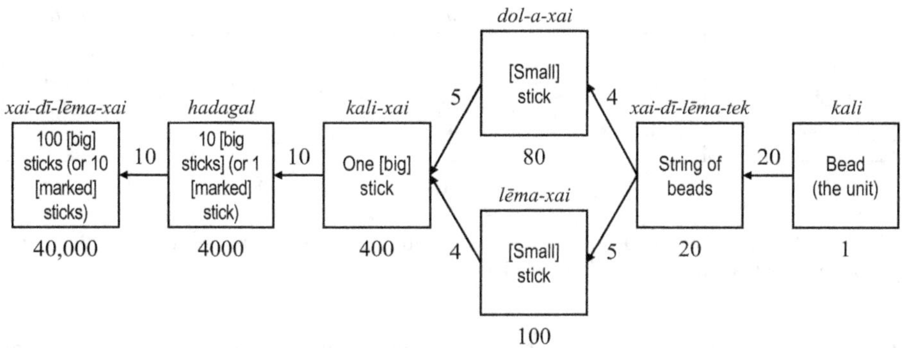

Fig. 33.2 Pomo counting units and organization. Image by the author, as diagramed from the vocabularies and descriptions in Barrett (1908), Dixon and Kroeber (1907), and Loeb (1926)

As shown in Fig. 33.2, each bead was worth *one*, or *kali* in the Eastern Pomo dialect.[7] Each string held *twenty* beads, *xai-dī-lēma-tek*, meaning "five full," a quinary-vigesimal construction[8] (that is, *twenty* is all *five* digits on all *four* limbs of a single person). *Four* (*dol*) strings of beads would be replaced by a small stick (*xai*) measuring about 1.75 inches in length and worth 80 beads (*dol-a-xai*); alternatively, a small stick might represent *five* (*lēma*) strings of beads, in which case it was worth 100 beads (*lēma-xai*).[9] When *five* small sticks collectively worth 80 beads (or *four* worth 100 beads) had been accumulated, they were replaced by *one* large stick measuring about 3.0 inches in length (*kali-xai*, worth 400 beads).[10] When *ten* of these large sticks had been accumulated, they might be replaced by a slightly larger stick with some distinguishing mark (*hadagal*, worth *four thousand* beads); when *ten* of the large, marked sticks had been accumulated, it made *forty thousand* beads, the "big twenty" called *xai-dī-lēma-xai*.[11]

The convention with *five* strings was perhaps intended to facilitate trade with neighboring tribes, whose number systems were often decimal. It might also have reflected an early Western interaction (a "Russian influence"[12]), given the decimality of Western numbers. The highest Pomo numbers are overly decimal in their organization, since the vigesimal powers above 20 and 400 are 8000 and 160,000, not 4000 and 40,000; this again suggests the influence of decimal-using neighbors and trading partners.[13] This kind of "mixed system" occurs frequently in California

[7] Barrett (1908).

[8] Dixon and Kroeber (1907), p. 685; also see Barrett (1908).

[9] Loeb (1926), p. 230.

[10] Loeb (1926), p. 230.

[11] Loeb (1926), p. 230.

[12] Farris (1990), p. 179.

[13] Kroeber (1911), p. 332.

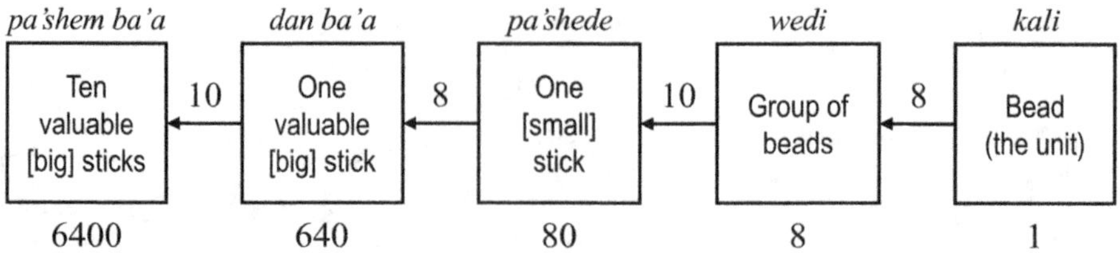

Fig. 33.3 Elem Pomo decimal-octal system for counting beads. Image by the author, as diagramed from the vocabulary and description in Ahlers (2012)

languages, "at least partly due to the borrowing of both numerals and bases": words for numbers and the principles for organizing them.[14]

Some external influences were octal, based on the number eight. One near neighbor with octal numbers was the Yuki people, whose distinctive octal numbers are discussed in the next chapter. Another Pomo tribe—the Elem to the southeast, whose territory was adjacent to the Yuki's southern territory[15]—had a fairly typical system of decimal numbers, along with an interesting decimal-octal counting sequence used with beads (Fig. 33.3).[16] A collective unit, a group of *eight* beads (*wedi*), formed the basis of the counting sequence, which could count into the tens of thousands. Some of the Elem number terms were subtractive in their formation: *Xot pa'shem wedi* (*seventy-two*) meant "[*eight*] less ten *wedi*" or (10 × 8) – 8, and *dan ba'a pa'shem Xai xut* (*five hundred and sixty*) meant "*one* valuable [stick] less *ten* [small] sticks" or (8 × 80) – 8.[17] The decimal-octal counting sequence is similar to West African cowrie-counting in that both use a group of beads as their basic (collective) unit. Its recurrent 8–10 pattern also evokes the recurrent 10–6 pattern of Sumerian numbers; since that 10–6 patterning is why Sumerian numbers are classified as sexagesimal (base 60), a number system with 8–10 patterning might be properly classified as octogesimal (base-80).

In addition to the sticks that represented bundles of strung beads, the Pomo used knotted strings to keep track of their bead exchanges with their neighbors. Each knot was equivalent to *four hundred* beads, and when completed, the knotted strings were "used as a check on the tying together of the counting sticks" accumulated during a counting session; the string and stick records were then "put away together in a bag and kept" until they were no longer needed, at which point they would be discarded.[18]

No trace of the Pomo number systems is found archaeologically. Pomo beads first appear in the archaeological record between 2500 and 500 years ago as artifacts

---

[14] Ahlers (2012), p. 537.

[15] Dixon and Kroeber (1907).

[16] Ahlers (2012).

[17] Ahlers (2012), p. 539.

[18] Loeb (1926), p. 231.

made of stone,[19] rather than the clamshell beads used as currency; evidence of clamshell use does not appear until much later—only within the past four or five centuries.[20] This recent date is consistent with the easy analyzability (or "etymological transparency") of the Pomo terms for higher numbers: *Dol-a-xai*, the term for *eighty*, meant *four sticks*, *lēma-xai* (*hundred*) meant *five sticks*, *kali-xai* (*four hundred*) meant *one [big] stick*, *hadagal* (*four thousand*) meant *ten [big sticks]*, and *xai-di-lēma-xai* (*forty thousand*) meant *one hundred [big] sticks*. As was noted for Intuit numbers, easy analyzability is generally indicative of recency of origin and/or infrequency of use; in this case, we do not suspect infrequent use, given the high extent of Pomo counting.

Sticks and knotted strings are perishable materials, and there are no diagnostic methods or criteria for discerning whether beads were counted in addition to being strung and perhaps worn. The Pomo don't appear to have recorded their numbers with more permanent media like bone or stone, so there is no trace of their highly elaborated number system in the archaeological record. What we know about their number system comes from vocabularies and descriptions of counting behaviors collected and recorded by Westerners, something that only occurred within the past century or so. Absent an Indigenous writing system, such evidence depends on contact with a society that has writing and is also motivated to use it to record its encounters with other societies.

Most of the prehistory of numbers falls into this same situation. Obviously, we can't collect linguistic and behavioral data on past societies, so we must depend on the material remains they leave behind. For numbers to remain in the material record, two conditions must be met: First, they must be recorded in a medium that preserves archaeologically, and second, they must have an unambiguous form, typically the presence of characteristic numerical grouping. As was discussed in the section on European numbers, linear marks on prehistoric bones meet the first criterion, but not the second. The Pomo numbers meet the second criterion but not the first, and thus, we are fortunate that our knowledge of this highly elaborated number system did not depend on archaeological evidence.

References

Ahlers, J. C. (2012). Two eights make sixteen beads: Historical and contemporary ethnography in language revitalization. *International Journal of American Linguistics, 78*(4), 533–555.

Barrett, S. A. (1908). *The ethno-geography of the Pomo and neighboring Indians*. University of California Press.

Basgall, M. E. (1982). Archaeology and linguistics: Pomoan prehistory as viewed from Northern Sonoma County, California. *Journal of California and Great Basin Anthropology, 4*(1), 3–22.

Dixon, R. B., & Kroeber, A. L. (1907). Numeral systems of the languages of California. *American Anthropologist, 9*(4), 663–690.

---

[19] Basgall (1982); Stewart (1985).

[20] Basgall (1982); Stewart (1985).

Farris, G. J. (1990). Vigesimal systems found in California Indian languages. *Journal of California and Great Basin Anthropology, 12*(2), 173–190.

Gorlinski, V. (2019). Pomo. In *Encyclopaedia Britannica.* https://www.britannica.com/topic/Pomo-people

Hudson, J. W. (1897). Pomo wampum makers: An Aboriginal double standard. *Overland Monthly, 30*(176), 101–108.

Kroeber, A. L. (1911). *The languages of the coast of California north of San Francisco*. University of California Press.

Kroeber, A. L. (1925). *Handbook of the Indians of California.* Bulletin 78. Smithsonian Institution, Bureau of American Ethnology. U.S. Government Printing Office.

Loeb, E. M. (1926). Pomo folkways. *University of California Publications in American Archaeology and Ethnology, 19*(2), 149–405.

McLendon, S., & Lowy, M. J. (1978). Eastern Pomo and Southeastern Pomo. In W. C. Sturtevant (Ed.), *Handbook of North American Indians* (Vol. 8, pp. 306–323). U.S. Smithsonian Institute.

MoneyMuseum. (2024). *Pomo and Chumash-Indians: Hawock shell money (wampum). Premonetary currencies in America.* MoneyMuseum. https://www.moneymuseum.com/en/coins?&id=2281

Stewart, S. B. (1985). *Time before time: Prehistory and archaeology in the Lake Sonoma area.* U.S. Army Corps of Engineers.

Chapter 34
Northern California: Number Systems Based on Fours and Eights

Abstract The number system of the Yuki, Americans native to northern California, was based on fours and eights. The Yuki number system is often described as based on counting the four spaces between the fingers, and there are number systems known to do this, as for example, the number system of the Aymara of Peru. In actuality, the Yuki counted sticks that were held in the spaces between the fingers. The Yuki words for numbers, along with a few historical documents that describe the use of sticks in pairs, suggest a hypothetical reconstruction of the stick configurations used. Most likely one hand held the sticks, while the other hand placed and reconfigured them. In the Yuki system, the terms *eight* and *sixteen* appear to have been productive, which would make these numbers an octal-hexadecimal system. California was home to other counting systems based on *four*, as recorded by early missionaries. These include the number systems of the Ventureño Chumash, a people indigenous to central-southern coastal California and adjacent islands, and the Salinan, a tribe whose ancestral homeland was in the central California coastal region.

Keywords Yuki · Octal · Ventureño Chumash · Salinan · Aymara

The Yuki—like the Pomo, their immediate neighbors to the south—are Americans native to northern California. The ancestral homeland of the Yuki stretched from Humboldt Bay (located about 250 miles north of San Francisco) to the upper Russian River Valley (the beginning of the Pomo homeland), with its "heart" being the Round Valley; the Yuki have inhabited this territory "since time began."[1] The Yuki language became extinct in 1983 with the death of its last-known speaker.[2] In most dialects, the number systems were quinary-decimal or quinary-vigesimal.[3] In the Round Valley language, however, the number system was organized by fours

[1] *Pre-History of Round Valley* (2024).

[2] Balodis (2016).

[3] Closs (1993); Kroeber (1911).

K. A. Overmann, *Cultural Number Systems*, Interdisciplinary Contributions to Archaeology, https://doi.org/10.1007/978-3-031-83383-0_34

and eights, using a method of counting widely described as counting the spaces between the fingers.[4]

The Yuki didn't actually count the spaces between their fingers. Admittedly, we don't know exactly how they counted because there aren't any knowledgeable cultural informants to tell us, nor any detailed historical descriptions. However, we do know the counting elements were not, as is widely believed, the spaces between the fingers. Rather, counting involved sticks held between the fingers in pairs.[5] One hand likely held the sticks while the other organized and moved them;[6] this seems more feasible than both hands holding sticks, an arrangement leaving neither hand free to arrange them. The importance of sticks to Yuki counting is unsurprising, given how critical they were to the counting systems of close neighbors like the Pomo.

The Yuki language had three "numeral roots" or terms for numbers that are unanalyzable ("etymologically opaque"); these were the words for *one*, *two*, and *three* (Table 34.1).[7] All other terms for numbers were "composite words descriptive of the method of counting."[8]

While these translations don't tell us exactly how the Yuki counted, we can speculate about the sticks and positions, as the number terms and some descriptive hints say that twigs were used in groups of *two* or fewer. Accordingly (Fig. 34.1), *four* might have been two pairs of twigs held between the index/middle and ring/little

Table 34.1 Round Valley Yuki numbers (compiled from Balodis, 2016, pp. 333–335; Kroeber, 1925, p. 177)

Nr.	Yuki	Meaning	Nr.	Yuki	Meaning
1	*pąwe*	(*one*)	11	*molmisul*	Three-body (3 + 8)
2	*ˀope*	(*two*)	12	*ˀomahątsul*	Two-forks-body (4 + 8)
3	*molme*	(*three*)	13	*huykosul*	Middle-in-body (5 + 8)
4	*ˀopmahąt*	Two-forks	14	*mikasčilkisul*	Even-*chilki*-body (6 + 8)
5	*huyk'o*	Middle-in	15	*mikaskosul*	Even-in-body (7 + 8)
6	(*mek'ąs*) *č'ilkeˀ*	Even-*chilki*	16	*huyčot*	Middle-none (8 + 8)
7	*mikasko*	Even-in	17	*pąwihuyluk*	One-middle-project (1 + 16)
8	*pawmpat*	One-flat	18	*ˀopihuyluk*	Two-middle-project (2 + 16)
9	*hučampąwipan*	Beyond-one-hang (1 + 8)	19	*molmihuy poy*	Three-middle-project (3 + 16)
10	*hučamopisul*	Beyond-two-body (2 + 8)	20	*ˀomahąthuypoy*	Two-forks-middle-project (4 + 16)

[4] Dixon and Kroeber (1907).

[5] Kroeber (1925), pp. 176–177.

[6] Kroeber (1925).

[7] Kroeber (1911).

[8] Kroeber (1911), p. 368.

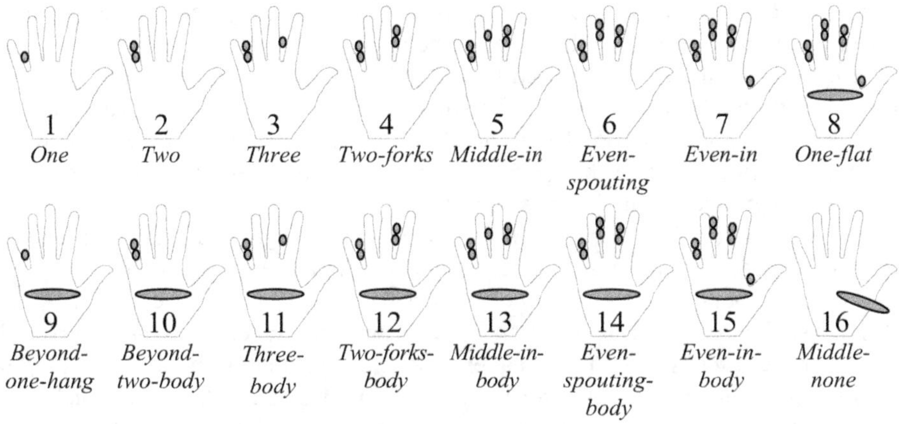

Fig. 34.1 Hypothetical reconstruction of finger-and-stick positions for the Yuki numbers *one* through *sixteen*, as seen from the counter's perspective with the right hand holding the sticks palm-up. Image by the author

fingers, an arrangement with two forks. This might make *five* the addition of a single twig in the middle of those forks. *Six* might then be the completion of the middle pair, at which point the four fingers are "spouting" twigs.[9] *Seven* might then be the insertion of a twig between the thumb and index finger. Finally, *eight* would be "two sticks between each two fingers";[10] alternatively, a single twig might be laid "flat" across the open palm. *Nine* through *fifteen* would then be the single twig held "flat" on the palm plus the sticks for *one* through *seven*. *Sixteen* might perhaps reposition the twig on the palm to make it "project" in some way from the hand.

Beyond *twenty*, most of the Yuki terms for numbers have been lost to time, so there's no way to know the extent of the Round Valley counting sequence, or how it might have compared to those of their neighbors. However, *forty* was *huyšot pawm-pat poy*, while *sixty-four* was *ʔomahąt šam op*.[11] The first means "two-middle-project-one-flat," or (2 × 16) + 8, while the second is "two-fork-pile-at,"[12] perhaps (4 × 16). These terms also hint that *sixteen*, like *eight*, was productive, which would make these numbers an octal-hexadecimal system.[13] Four other higher numbers are known only through cryptic clues: "24 = 8, 26 = 10, 35 = 19, 51 = 19."[14] The clues for 24 and 26 suggest the numbers 17 through 31 were built on 16: 24 = 8 + 16 and 26 = 10 + 16. The clue for 51, plus the term for 40, suggest 33–63 were built on 32: 51 = (2 × 16) + 19. Not fitting this pattern is the clue for 35, as we would predict

[9] Sawyer and Schlichter (1984), p. 189.

[10] Sawyer and Schlichter (1984), p. 76.

[11] Balodis (2016).

[12] Kroeber (1925), p. 177.

[13] Closs (1993).

[14] Dixon and Kroeber (1907), p. 684.

(2 × 16) + 3 and not 19 + 16. Indeed, number systems don't typically count beyond the grouping number (16, in this case) in the manner suggested.

The sticks could have been used to add and subtract in a manner similar to one described for the Ryukyu Islands, the chain of islands that extend from Kyushu (the southernmost large island of Japan) and Taiwan. There, stalks of rush straw were inserted or removed from the spaces between the fingers, and then simplified, as required, to produce numerical answers.[15]

Some societies might count the spaces between the fingers. The Aymara (or Aimara), a people indigenous to the Altiplano and Andes regions of Peru, have been described as "not only count[ing] the digits, but also the spaces between them."[16] As Aymara spoken numbers are decimal,[17] and because *five* fingers with *four* interstitial spaces make *nine*, the counting sequence would presumably include an external position (or some other convention) to permit counting to reach *ten* per hand. The brief (and dated) description of Aymara counting has no hint of using sticks as the counted elements, so if they do count the spaces along with the fingers, they do something that is extremely rare—not even the Yuki did it. In any case, the Aymara are far enough away from the Yuki to make it highly unlikely the two counting practices were connected in any way.

Northern California was home to other counting systems based on four. One of these belonged to the Ventureño Chumash, a people indigenous to central-southern coastal California and adjacent islands. These numbers were initially recorded in a "'confesionario' or handbook for the confessor," identified by its handwriting as written by Father José Señán, a Franciscan monk at the San Buenaventura Mission, most likely between 1812 and 1815;[18] this makes these Californian numbers some of the earliest documented.[19] Señán's account, given modern spellings (Table 34.2), shows unanalyzable terms for the numbers *one* through *three*; thereafter, the terms are based on multiples of *four* and *sixteen*, with a mix of additive and subtractive forms. Señán recorded two higher numbers: *Ninety-six* was *yəti'iškoṁ tšikipš* or (6 × 16), and *one hundred twenty-eight* was *malawa tšikipš*, (8 × 16).[20] The reason Chumash numbers were octal is unknown; the last cultural informant fluent in the language was said not to know about any involvement of sticks.[21]

A third Californian number system organized by fours (Table 34.3) belonged to the Salinan, a tribe whose ancestral homeland was in the central California coastal region. These numbers were originally recorded by Buenaventura Sitjar, a Franciscan monk at the San Antonio Mission, but not published until 1861, more than 50 years after his death.[22] Interestingly, the formations for *five* and its multiples are related to

[15] Simon (1924).

[16] Bennett (1949), p. 614.

[17] McGourn (1990).

[18] Beeler (1964), p. 16; also see Señán (1967), p. 4.

[19] Beeler (1986); also see Hughes (1974).

[20] Beeler (1986).

[21] Beeler (1986).

[22] Sitjar (1861).

Table 34.2 Ventureño Chumash numbers (Beeler, 1986; also see Beeler, 1964; Señán, 1967)

Nr.	Chumash	Meaning	Nr.	Chumash	Meaning
1	*pake'et*	1	17	*tšikipš kampake'et*	(4 × 4) + 1
2	*'iškoṁ*	2	18	*'iškoṁ siwe tskumu'uy*	2 from (5 × 4)
3	*masəx*	3	19	*pake'et siwe tskumu'uy*	1 from (5 × 4)
4	*tskumu*	4	20	*tskumu'uy*	(5 × 4)
5	*yətipake'es*	4 + 1	21	*tskumu'uy kampake'et*	(5 × 4) + 1
6	*yəti'iškoṁ*	4 + 2	22	*'iškoṁ siwe itsmaxmasəx*	2 from (6 × 4)
7	*yətimasəx*	4 + 3	23	*pake'et siwe itsmaxmasəx*	1 from (6 × 4)
8	*malawa*	8	24	*itsmaxmasəx*	([twice 3] × 4)
9	*tspa*	9	25	*itsmaxmasəx kampake'et*	(6 × 4) + 1
10	*ka'aškom*	8 + 2	26	*'iškoṁ siwe yitimasəx*	2 from (7 × 4)
11	*təlu*	11	27	*pake'et siwe yitimasəx*	1 from (7 × 4)
12	*masəx tskumu*	(3 × 4)	28	*yitimasəx*	(7 × 4)
13	*masəx tskumu kampake'et*	(3 × 4) + 1	29	*yitimasəx kampake'et*	(7 × 4) + 1
14	*'iškoṁ laliet*	2 from (4 × 4)	30	*'iškoṁ siwe 'iškoṁ tšikipš*	2 from (2 × 16)
15	*pake'et siwe (tšikipš)*	1 from (4 × 4)	31	*pake'et siwe 'iškoṁ tšikipš*	1 from (2 × 16)
16	*tšikipš*	(4 × 4)	32	*'iškoṁ tšikipš*	(2 × 16)

Table 34.3 Salinan numbers (Mason, 1912, p. 136; also see Turner, 1987)

Nr.	Salinan	Meaning	Nr.	Salinan	Meaning
1	*t'ol*	All alone	9	*t'o'l-te-ṭ'o'‘e*	One-from-[plural]-fists (10 – 1)
2	*ka'k'-ca'*	Half-four (4 ÷ 2)	10	*ṭ'o'‘e*	[two] fists (2 × 5)
3	*la'pai*	Three	11	*ṭ'o'‘e-tax-t'ol*	[two] fists and one (10 + 1)
4	*k'ca'*	It-is-four	12	*lapai-k'ca'*	Three-it-is-four (3 × 4)
5	*t'o'lṭ'a'o*	One-his-fist (1 × 5)	13	*lapai-k'ca'-tax-t'ol*	Three-it-is-four-and-one (12 + 1)
6	*lapai-ā'nel*	Threes [plural] (2 × 3)	14	*wococo*	Fourteen
7	*te'*	Seven	15	*lapai-(t')ol-ṭ'a'o*	Three-one-his-fist (3 × 5)
8	*ca'-a'nel*	Fours [plural] (2 × 4)	16	*k'pec*	It-is-sixteen

the word for "fist," suggesting the use of the fingers in counting. While the reason why this number system was organized by fours is unknown, the use of *five* and its multiples suggests a typical finger-counting system either in contact with the practice of counting by fours or adopting a technology (like the Yuki sticks) that influenced its development.

References

Balodis, U. (2016). *Yuki grammar: With sketches of Huchnom and Coast Yuki*. University of California Press.

Beeler, M. S. (1964). Ventureño numerals. *Studies in Californian Linguistics, 34*, 13–18.

Beeler, M. S. (1986). Chumashan numerals. In M. P. Closs (Ed.), *Native American mathematics* (pp. 109–128). University of Texas Press.

Bennett, W. C. (1949). Mnemonic and recording devices. In J. H. Steward (Ed.), *Handbook of South American Indians. Vol. 5: Comparative ethnology of the South American Indians* (pp. 611–619). U.S. Government.

Closs, M. P. (1993). Native American number systems. In M. P. Closs (Ed.), *Native American mathematics* (pp. 3–43). University of Texas Press.

Dixon, R. B., & Kroeber, A. L. (1907). Numeral systems of the languages of California. *American Anthropologist, 9*(4), 663–690.

Hughes, B. B. (1974). The earliest known record of California Indian numbers. *Historia Mathematica, 1*(1), 79–82.

Kroeber, A. L. (1911). *The languages of the coast of California north of San Francisco*. University of California Press.

Kroeber, A. L. (1925). *Handbook of the Indians of California*. Bulletin 78. Smithsonian Institution, Bureau of American Ethnology. U.S. Government Printing Office.

Mason, J. A. (1912). *The ethnology of the Salinan Indians*. University of California Press.

McGourn, F. T. (1990). Aymara, Bolivia, Chile, Argentina. In E. Chan (Ed.), *Numeral systems of the world's languages*. Max Planck Institute. https://lingweb.eva.mpg.de/channumerals/Aymara-Central.htm

Pre-history of Round Valley. (2024). *Tribal territory since time began: A sovereign nation of confederated tribes*. Round Valley Indian Tribes. https://www.rvit.org/about/about-us

Sawyer, J. O., & Schlichter, A. (1984). *Yuki vocabulary*. University of California Press.

Señán, J. F. d. P. (1967). *The Ventureño confesionario of José Señán, O.F.M.* University of California Press.

Simon, E. (1924). Über Knotenschriften und ähnliche Knotenschnüre der Riukiu-Inseln [On knot writing and similar knotted cords of the Riukiu Islands]. *Asia Major, II*, 657–667.

Sitjar, B. (1861). *Vocabulario de la lengua de los naturales de la Mission de San Antonio, Alta California [Vocabulary of the language of the natives of the Mission of San Antonio, Alta California]*. Cramoisy Press.

Turner, K. (1987). *Aspects of Salinan grammar.* Unpublished PhD dissertation, University of California at Berkeley.

Chapter 35
Mesoamerica: Maya Numbers

Abstract Maya numbers were quinary-vigesimal, and the associated notation systems were positional. One system consisted of bars and dots. Positionality was maintained by a sign resembling a seashell. A second notational system consisted of glyphs that depicted leftward-facing heads. The earliest information about Maya numbers comes from Diego de Landa, the Conquistador who infamously burned all the Maya books he could find. De Landa described the Maya number system as reaching infinity, though the highest term meant *twenty* to the seventh power, or 12.8 billion. Maya calculating involved counters placed on a flat surface. In the 1970s, the *Nepohualtzintzin*, a vigesimal abacus with 13 columns of seven beads, was created from purported oral histories and artifactual depictions. This device represents a significant revitalization of the Mesoamerican vigesimal number systems and their complex mathematical traditions. Today, the seashell sign used as a placeholder is often called *zero*, but conceptually, it was unlikely to have had the same meaning that *zero* does in today's Western system. Given its calendrical use, the Maya sign might have meant "completion," rather than "nothing."

Keywords Quinary-vigesimal · Positional numbers · Diego de Landa · *Nepohualtzintzin* · *Zero*

Mesoamerica, the middle ground connecting the north and south American continents, was home to several pre-Hispanic civilizations: the Olmecs, Maya, Toltecs, and Aztecs. Of these, Maya are a civilization of peoples inhabiting a territory centered on the Yucatán Peninsula, a region that today includes the lowlands of southern Mexico, most of Guatemala and Belize, and the western portions of Honduras and El Salvador.[1] By 800 CE, the Maya Empire was at its height, with achievements that included agriculture, a system of writing, and great stone cities that could house up to 100,000 people; by just a century later, the Maya had abandoned most of their cities, a rapid political and social collapse attributed to environmental degradation

[1] Closs (1993).

K. A. Overmann, *Cultural Number Systems*, Interdisciplinary Contributions to Archaeology, https://doi.org/10.1007/978-3-031-83383-0_35

and drought.[2] Today, the Maya are known for their ornate temples, complex astronomy, and a calendar famously misunderstood as predicting the world would end in 2012. They also had two systems for writing numbers—bar-and-dot notations and head glyphs—both of which included a sign for *zero*.

Maya numbers were quinary-vigesimal, and their notation systems were positional. The quinary-vigesimal structure is easily seen in the bar-and-dot numerals (Fig. 35.1, left), where signs are organized in four groups of 5 by the number of bars, and 20 is productive. Positionality is maintained by a placeholder, a sign called *zero* that resembles the long, oval shells of Oliva sea snails.[3] Signs are stacked to represent the different multiples of 20, with the notations on the bottom being the units. Thus, the sign for 20 is the sign for 1 placed above the sign for 0, the sign for 400 (20 squared) is the sign for 1 placed above two stacked 0, and the sign for 8000 (20 cubed) is the sign for 1 placed above three stacked 0; these stacks are the vigesimal analogues of 10, 100, and 1000 in a decimal system. As is true of the Kaktovik numerals of the Inuit, addition and subtraction with the Maya bar-and-dot numerals involves consolidating or removing the various elements.[4]

The bars may have been intended to resemble cacao pods.[5] The seeds of these pods are the source of chocolate, which was used as a beverage, spice, medicine, and aphrodisiac, with pods and seeds also serving as ritual objects and currency. The terms for higher numbers in some Maya languages include names suggesting cacao bundles. For example, in Kaqchikel, a language spoken in Guatemala, the word *chuwi* (*eight thousand*) means "sack"; "its use as a numeral is said to derive from the custom of packaging cacao beans—an important commodity and also a medium

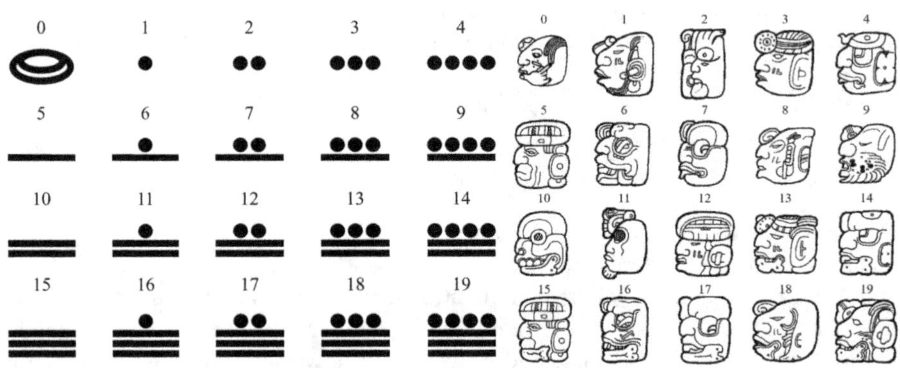

Fig. 35.1 (Left) Maya bar-and-dot numerical notations. Each bar has the value *five*, and each dot means *one*. Image by the author. (Right) Maya head glyph numerical notations. Adapted from an image in the public domain

[2] Kennett et al. (2012); Penuelas and Nogué (2023).

[3] Blume (2011).

[4] Lambert et al. (1980).

[5] Thompson (1941).

of exchange—in quantities of 8000 to the bag."[6] However, the Maya were unlikely to have used the cacao pods themselves for calculating: These football-shaped objects typically measure 8–14 inches in length,[7] a size meaning that even simple calculations would require extremely large boards! Perhaps the Maya, like the American peoples to their north and south, used sticks and kernels of corn to represent the bars and dots, with the cacao pods and seeds serving to inspire their artistic forms.

Numerals could also be written with head glyphs. The numeral glyphs depict the leftward-facing heads of gods and mortals, as distinguished by their facial features, hairstyles, headdresses, and jewelry (Fig. 35.1, right). For example, the glyph for *seven* is the head of one of several Maya Jaguar Gods, identified by the curl in his eye and his sharp front tooth. These signs were part of a larger system of writing capable of representing both the meanings and sounds of several of the languages spoken by different Maya peoples.[8]

Our earliest look at Maya numbers comes from the man who destroyed their history. Diego de Landa (1524–1579) was a bishop and inquisitor during the Conquest, which for the Yucatán Peninsula began in 1517 with the first Spanish landfall and ended in 1697 with the defeat of the last independent Maya kingdom.[9] De Landa's historical significance comes from two things: First, he wrote a book about the Yucatán in 1566, though this account would not be published for another three centuries; second, he burned all the Maya books he could lay his hands on in 1562.[10] This "[in]famous Auto de fé" destroyed "ninety-nine times as much knowledge of Maya history and sciences as he has given us in his book."[11] De Landa briefly described the Maya number system as follows: "They count by fives up to twenty, by twenties to a hundred and by hundreds to four hundred; then by 400s up to 8000. This count is much used in merchandising the cacao. They have other very long counts, extended to infinity, counting twenty times 8000, or 160,000; then they multiply this 160,000 again by 20, and so on until they reach an uncountable figure. They do all their counting on the ground or a flat surface."[12]

Although de Landa said Maya numbers reached "*infinitum*,"[13] the highest calendrical term was *alau*, meaning 64 million (Fig. 35.2).[14] There was also a term for the seventh power of 20, *hablat*, meaning 12.8 billion.[15] An upper limit of the seventh power is consistent with the extents of many ancient number systems, as well as one

[6] Closs (1993), pp. 293–294; also see Brinton (1882).

[7] *Cacao Trees and Pods* (2024).

[8] Houston et al. (2000).

[9] Sharer and Traxler (2006).

[10] Gates (1978).

[11] Gates (1978), pp. 9–10.

[12] De Landa (1864), p. 134, as translated in De Landa (1978), p. 62.

[13] De Landa (1864), p. 134, emphasis in original.

[14] De Landa (1978), editor's note on p. 62; also see Dorado (2002), p. 88.

[15] *The Mayan Numeral System* (2014).

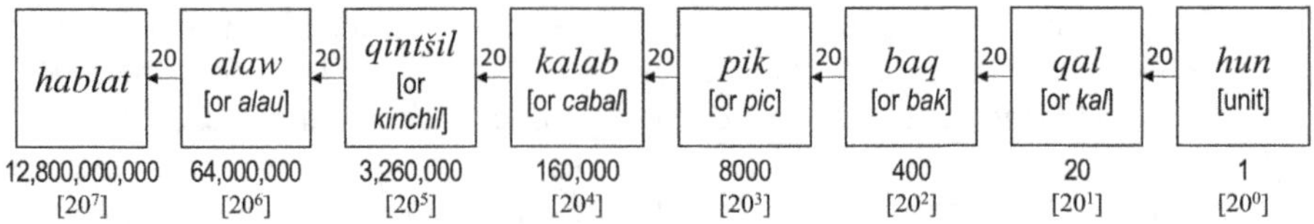

Fig. 35.2 The Maya number system. Image by the author, as diagramed from the terms and values in Tozzer (1921, p. 97) and *The Mayan Numeral System* (2014)

of the highest known in the Americas. However, while it exceeds the counting range of the New England Narragansett (Wampanoag or Massachusett) at 100,000 (the fifth power of *ten*),[16] it is dwarfed by the Inka at *ten hundred thousand million million* (a quintillion, the eighteenth power of *ten*; see Table 36.1).[17]

Maya calculating involved counters on a flat surface. If the Maya had a counting board of some sort, we don't know what it might have looked like, nor indeed, many details of how calculations were performed.[18] De Landa said the Maya used the ground or other flat surface for calculating[19]; this implies a level space without any partitions or markings—nothing as organized or formalized as an abacus.[20]

In the 1970s, a vigesimal abacus was created from descriptions in oral histories and depictions on bracelets and scenes suggestive of accounting (Hidalgo, 1977). It was named *Nepohualtzintzin* (pronounced NEH-poh-hoo-wahl-TZIN-tzin; often shortened to "Nepo"), a term in the Nahuatl language translated variously as *nepohual* ("counting") and *tzintzin* ("venerable or relevant")[21]; *ne-* ("personal"), *pohual* ("account"), and *tzintzin* ("small elements")[22]; *ne* ("person"), *pohual* ("to count or to tell a story"), and *tzintzin* ("to transcend or to share what one knows")[23]; or "cord of counting and number," a name derived from "Nepohuàlco, which means Contadèro" [Accountant].[24] In accounts from the eighteenth and nineteenth centuries, the Nepo was understood not as a calculating device, but rather, as a Mexican version of the *khipu*, the stringed device used by the Inka to record their numbers; moreover, none of the authors appears to have seen an intact specimen of an historical Nepo for himself.[25]

As recreated, the Nepo has 13 columns of seven beads (Fig. 35.3). In each column, the three top beads represent fives, while the four on the bottom are ones; beads are counted when they are moved toward the Nepo centerline. *Twenty* in any

16 Williams (1643).

17 Gonçalez Holguin (1607), Book III, Ch. 52, p. 100.

18 Freidel and Rich (2015); Long (1948).

19 De Landa (1864).

20 Long (1948).

21 Orbe (2024).

22 García Maciel (2009).

23 Casey (2024).

24 Boturini Benaduci (1746), pp. 85–87.

25 Boturini Benaduci (1746); Von Humboldt (1814).

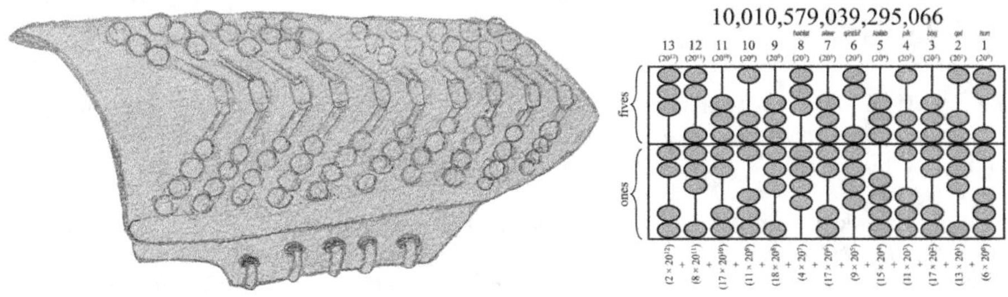

Fig. 35.3 The *Nepohualtzintzin.* (Left) As Aztec bracelet with ten columns. Drawing by Charles Payne based on Hidalgo (1977, plate 42). (Right) Shown as a schematic diagram with 13 columns. Image by the author

column shifts 1 to the next higher column and returns all the original beads to their starting positions. The 13 columns count to extremely high numbers; in Western notation, 19 of the highest unit (20^{12}) is 81,919,999,999,999,999, somewhat more (in fact, 81,919,998,719,999,999 more) than the 12.8 billion that is the seventh power of 20. This capacity is consistent with a modern origin, not an ancient one, as ancient peoples seldom required numbers much beyond a million. Even modern peoples seldom require such large numbers; the number is so big, it exceeds the capacity of Excel, the Microsoft spreadsheet software. The Nepo nonetheless represents a significant revitalization of the Mesoamerican vigesimal number systems and their complex mathematical traditions.

Did the Maya really have a zero*? It depends on how* zero *is defined. Zero* was already old by the time the Maya were using it: Its earliest representation, which is associated with the Olmec culture, dates to 31 BCE.[26] Because they used counting boards (however loosely defined) with counters, the Maya didn't calculate with written notations and the kind of algorithms found today in the West; in both, *zero* is a placeholder, but in the Western tradition it has also become a number in its own right (a conceptualization enabled by its being represented by a numeral,[27] a form enabling it to be manipulated just like any other number; in comparison, the *zero* implied by abacus bead positions cannot be similarly manipulated). This Western conceptualization is "fairly recent," so it's "extremely unlikely" the Maya had the same idea.[28] But the Maya *zero* might also have meant more than simply "no number goes here": It may have "represented completion rather than 'nothing'"; within the context of the Maya calendar, this might be understood as "cyclical completion, … the same as a new beginning."[29]

---

[26] Batz (2011), p. 1.

[27] Otis (2024).

[28] Fulton (1948), p. 1.

[29] Fulton (1948), p. 3.

References

Batz, J. M. (2011). "Nik"—The zero in vigesimal Maya mathematics. *Bulletin of the American Astronomical Society, 53*(1), 1–5.

Blume, A. (2011). Maya concepts of zero. *Proceedings of the American Philosophical Society, 155*(1), 51–88.

Boturini Benaduci, L. (1746). *Idea de una nueva historia general de la America septentrional [Idea of a new general history of North America]*. Juan de Zuñiga.

Brinton, D. G. (1882). *Library of Aboriginal American literature. Vol. 1: The Maya chronicles*. Daniel G. Brinton.

Cacao Trees and Pods. (2024). *The story of chocolate*. National Confectioners Association. https://candyusa.com/story-of-chocolate/where-does-chocolate-come-from/cacao-pods/

Casey, M. (2024). *Educators want Mesoamerican math tool Nepohualtzintzin in Arizona classrooms*. Fronteras. https://fronterasdesk.org/content/1463956/educators-want-mesoamerican-math-tool-nepohualtzintzin-arizona-classrooms

Closs, M. P. (1993). The mathematical notation of the ancient Maya. In M. P. Closs (Ed.), *Native American mathematics* (pp. 291–369). University of Texas Press.

De Landa, D. (1864). *Relation des choses de Yucatan de Diego de Landa [Diego de Landa's account of things in Yucatan]*. Auguste Durand.

De Landa, D. (1978). *Yucatan before and after the Conquest* (W. Gates, Ed. & Trans.). Dover Publications.

Dorado, M. R. (Ed.). (2002). *Relación de las cosas de Yucatán [List of things from Yucatan]*. Dastin.

Freidel, D. A., & Rich, M. (2015). Pecked circles and divining boards: Calculating instruments in ancient Mesoamerica. In A. S. Dowd & S. Milbrath (Eds.), *Cosmology, calendars, and horizon-based astronomy in ancient Mesoamerica* (pp. 249–264). University Press of Colorado.

Fulton, C. C. (1948). Did the Maya have a zero? The meanings of our zero and the Maya "zero." *Notes on Middle American Archaeology and Ethnology, 90*, 1–3.

García Maciel, O. (2009). *Nepohualtzintzin*. Ábaco Azteca. http://abacoazteca.ucoz.com/

Gates, W. (1978). Introduction. In W. Gates (Ed.), *Yucatan before and after the Conquest* (pp. 9–21). Dover Publications.

Gonçalez Holguin, D. (1607). *Gramatica y arte dveva de la lengva general de todo el Peru, llamada lengua Qquichua, o lengua del Inca [Grammar and art of the general language of all Peru, called the Quichua language, or the language of the Inca]*. Francisco del Canto.

Hidalgo, D. E. (1977). *Nepohualtzintzin: Computador prehispánico en vigencia [Nepohualtzintzin: Pre-Hispanic computer in use]*. Editorial Diana.

Houston, S., Robertson, J., & Stuart, D. (2000). The language of Classic Maya inscriptions. *Current Anthropology, 41*(3), 321–356.

Kennett, D. J., Breitenbach, S. F. M., Aquino, V. V., et al. (2012). Development and disintegration of Maya political systems in response to climate change. *Science, 338*(6108), 788–791.

Lambert, J. B., Ownbey-McLaughlin, B., & McLaughlin, C. D. (1980). Maya arithmetic. *American Scientist, 68*(3), 249–255.

Long, R. C. E. (1948). Some remarks on Maya arithmetic. *Notes on Middle American Archaeology and Ethnology, 88*, 350–351.

Orbe, G. G. C. (2024). The Mesoamerican abacus that gives modern calculators a run for their money (G. G. Calderón, Trans.). *Global Voices*. https://globalvoices.org/2015/09/04/the-mesoamerican-abacus-that-gives-modern-calculators-a-run-for-their-money/

Otis, J. M. (2024). *By the numbers: Numeracy, religion and the quantitative transformation of early modern England*. Oxford University Press.

Penuelas, J., & Nogué, S. (2023). Catastrophic climate change and the collapse of human societies. *National Science Review, 10*(6), 1–3.

Sharer, R. J., & Traxler, L. P. (2006). *The ancient Maya* (6th ed.). Stanford University Press.

The Mayan numeral system. (2014). *Historical counting systems*. Northern Kentucky University. https://www.nku.edu/~longa/classes/2014fall/mat115/mat115-006/images/mayan/MayanCounting1.0.pdf

Thompson, J. E. S. (1941). *Maya arithmetic*. Carnegie Institution.

Tozzer, A. M. (1921). *A Maya grammar.* Peabody Museum of American Archaeology and Ethnology.

Von Humboldt, A. (1814). *Researches, concerning the institutions & monuments of the ancient inhabitants of America, with descriptions and views of some of the most striking scenes in the Cordilleras* (H. M. Williams, Trans.). Longman, Hurst, Rees, Orme & Brown, J. Murray, & H. Colburn.

Williams, R. (1643). *A key into the language of America: Or, an help to the language of the natives in that part of America, called New-England*. Gregory Dexter.

Chapter 36
Western South America: Inka numbers

Abstract The Inka "wrote" by tying knots in strings. Bundles of knotted strings are called *khipus*; these typically recorded decimal numbers in the thousands. Inka number-words up to *ten* to the eighteenth power, a quintillion, were documented in 1607 by a Jesuit missionary. Inka numbers are widely considered to be positional with place value and a *zero* (a blank section of string), but this interpretation may reflect familiarity with Western numbers. Inka numbers were more likely additive, organized like the Roman numerals. No chronicler described Inka calculations beyond noting that kernels of maize were used as counters, and these were moved with speed and skill. Three devices have been proposed for the *yupana*, the Inka calculating device. The first is a small, portable artifact with multiple levels and partitions, though it is likely from Ecuador and had uses other than numbers. The second is a checkerboard drawn in 1615; it has been subject to multiple mathematical analyses, with no agreed-upon interpretation. The third is the Inka storehouses whose stone floors are pitted with regularly spaced holes, though their size, depth, and spacing would make them difficult to use with counters.

Keywords *khipu* · Positional numbers · Additive numbers · Guamán Poma · *yupana*

Early in the thirteenth century CE, the Inka Empire was just beginning to form in the highlands of Peru.[1] Only three centuries later, it was the largest political entity in pre-Columbian South America, spanning the western portion of the continent. Spanish forces reached the Inka Empire at its height in 1528, won their first significant military victory over it in 1532, and completed their conquest in 1572. What the Spanish found astonished them.

The Inka "wrote" with string. As reported in the "chronicles," the literature of the discovery and conquest of the New World, the Inka emperor ruled an immense, complex state managed by four regional governors, dozens of regional

[1] Brotherston (1995).

K. A. Overmann, *Cultural Number Systems*, Interdisciplinary Contributions to Archaeology, https://doi.org/10.1007/978-3-031-83383-0_36

administrators, and hundreds of local administrators. Officials at each level collected and reported detailed demographic and economic data. This information was encoded, compiled, and sent to the highest levels of Inka authority—by means of knotted strings (Fig. 36.1, left). As a Spanish administrator noted in 1561, he wouldn't have believed string could be used as writing if he hadn't seen it for himself.[2]

Bundles of knotted strings are known as *khipus* (*quipus* or *quipos* in Spanish) after the word for "knot" in Quechuan, the Inka language. As described in 1590 by the Spanish missionary José de Acosta, "*Quipus* are memory aids or registers made up of cords on which different knots and different colors signify different things. What [the Inka] achieved in this way is incredible, for whatever books can tell of histories and laws and ceremonies and accounts of business all is supplied by the *quipus* so accurately that the result is astonishing."[3] The non-numerical component of *khipus* remains undeciphered; these knots were memory prompts whose meaning has been lost.[4] The numerical component was decoded in 1912 from a 1609 description by Garcilaso de la Vega, a Spanish soldier.[5] Three simple knots were used: Figure-eight knots and long knots with multiple twists were tied at the ends of

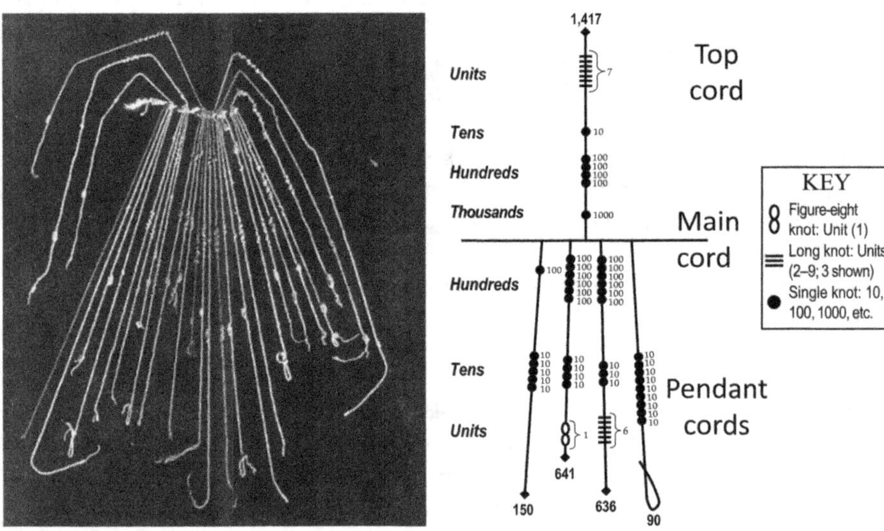

Fig. 36.1 *Khipu* B8713. (Left) Photograph by Locke (1912, Pl. XXIV). Image in the public domain. (Right) Portion of Locke's diagram of B8713 (*b*, second group from left). Redrawn and adapted from Locke (1912, pl. XXIII); key after Ascher and Ascher (1981). Note that Locke used additive notation in labeling the knots

[2] Polo de Ondegardo, as translated by Brokaw (2010).

[3] De Acosta (1590), as translated by Martin and Wasserman (2010), p. 5, italics added.

[4] Hyland (2020); Hyland et al. (2021); Medrano and Urton (2018); Urton (2010, 2017, 2018).

[5] Locke (1912).

strings to express the units *one* and *two* through *nine*, respectively; single knots tied above them were each valued as *ten*, *hundred*, or *thousand*, depending on their height on the cord (Fig. 36.1, right). Top cords displayed the sum of their pendant cords. *Khipus* could be large and complex: One (KH0082) has a primary cord measuring 16.8 feet (5.135 meters) that contains over 1800 cords, though the majority (78.2%) were unknotted.[6]

Inka numbers were decimal. Khipus typically recorded numbers in the thousands, infrequently, the tens of thousands,[7] and "seldom if ever passed a hundred thousand."[8] No known *khipu* represents a million. The Inka words for numbers (Table 36.1) were first documented in 1607 by Diego Gonçalez Holguín, a Jesuit missionary.[9] He described powers of *ten* up to a *quintillion*, the eighteenth power.[10]

Inka numbers are widely—but probably incorrectly—considered positional. A positional system has place value and *zero*. Numbers without place value or *zero* are called "additive." When additive numbers like the Roman numerals are written,

Table 36.1 Inka numbers (Gonçalez Holguin, 1607).

Quechuan terms	English translations	Decimal values
hue	One	1
chunca	Ten	10
pachac	One hundred	100
huaranca	One thousand	1,000
chunca huaranca	Ten thousand	10,000
pachac huaranca	Hundred thousand	100,000
hunu	One million	1,000,000
chunca hunu	Ten million	10,000,000
pachac huno	Hundred million	100,000,000
huaranca hunu	One thousand million	1,000,000,000
[chunca huranca hunu]	Ten thousand million	10,000,000,000
pachac huaranca hunu	One hundred thousand million	100,000,000,000
hunuy hunu	One million million	1,000,000,000,000
chunca hunuy hunu	Ten million million	10,000,000,000,000
pachac hunuy hunu	One hundred million million	100,000,000,000,000
huaranca hunuy hunu	One thousand million million	1,000,000,000,000,000
[chunca huaranca hunuy hunu]	Ten thousand million million	10,000,000,000,000,000
pachac huaranca hunuy hunu	One hundred thousand million million	100,000,000,000,000,000
chunca pachac huaranca hunuy hunu	Ten hundred thousand million million	1,000,000,000,000,000,000

Note: Items in parentheses were not listed in the original

[6] Ascher and Ascher (1981).

[7] Ascher and Ascher (1981); Chrisomalis (2010).

[8] De la Vega (1609), as translated in Livermore (1966), pp. 330–331.

[9] Gonçalez Holguin (1607).

[10] Gonçalez Holguin (1607), Book III, Ch. 52, p.100.

character form encodes value: *one* (I), *ten* (X), *hundred* (C). Since Inka numbers were not written, knots were valued as 1, 2–9, or 10 and higher by the type of knot used and where they were tied along the string. A blank stretch of string meant the absence of any number with that value. Since positional and additive numbers yield the identical knots (Fig. 36.2), Western scholars likely interpreted the *khipu* as the system they were more familiar with.

How the Inka calculated is a mystery. No chronicler described it in any detail, beyond noting the Inka used grains or pebbles as numerical counters and moved them with great skill. The grains were *maize* (corn), not *quinoa* as is sometime supposed, as quinoa is too small to be easily grasped or viewed. Maize comes in many colors, and color was likely used to express the different decimal values.[11] Using maize was not unique, as the New England Narragansett (Wampanoag) were described as using kernels of corn for calculating at about the same period of time.[12] Like the Narragansett, the Inka were also highly skilled at calculating. As described by de Acosta in 1590, "[The Inka] will take their grains and put one here, three there, and eight I don't know where."[13] In 1609, de la Vega observed, "[The Inka] added, subtracted, and multiplied with these knots, and ascertained the dues of each town by dividing grains of maize and pebbles."[14] Although de la Vega suggested

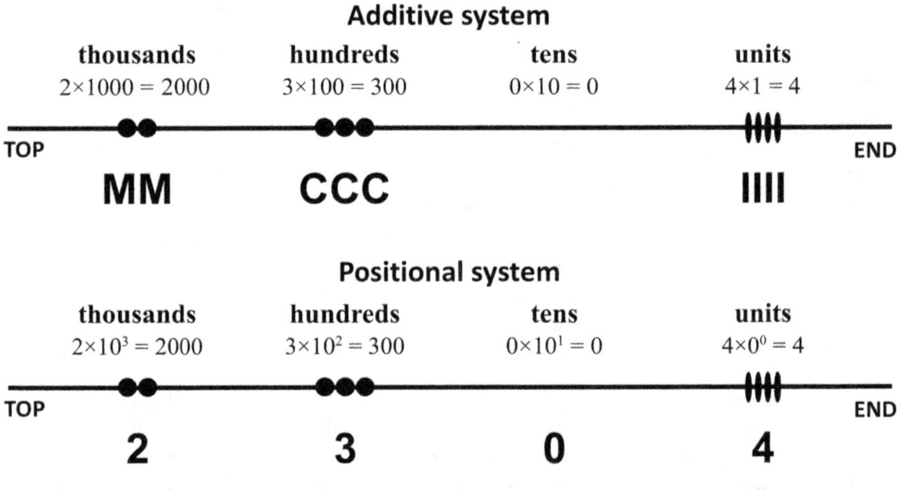

Fig. 36.2 (Top) In an "additive" system like the Roman numerals, each sign encodes a specific value that is then added to produce the total value. (Bottom) In a "positional" system like the Western numerals, signs take part of their value from their place, rather than encoding everything with their form. While the same *khipu* knots support both interpretations, the additive interpretation is less likely to reflect familiarity bias on the part of Western scholars. Image by the author. A version was previously published in Overmann (2023, Fig. 12.4, p. 287)

[11] Florio (2009).

[12] Williams (1643).

[13] De Acosta (1590), as translated by Martin and Wasserman (2010), pp. 5–6.

[14] De la Vega (1609), as translated by Livermore (1966), p. 124.

khipu knots were used for many arithmetic operations, knots in strings cannot be tied, untied, and retied fast enough for this purpose; perhaps these operations were performed mentally and then recorded with string.

In de la Vega's account, division, at least, involved rearranging numerical counters. The surface used was called a *yupana*, from the Quechuan word *yupay*, meaning "to count." Together, the *yupana* and its counters made a kind of abacus. Since no chronicler described a *yupana*, we don't know what one looked like or how it worked. One possibility is a small (portable) artifact with multiple levels and partitions (Fig. 36.3, left). In his 1789 chronicle, the Jesuit historian Juan de Velasco mentions a calculating device with multiple partitions; however, he didn't note any different levels, and he said the device belonged to the culture of Quito (modern Ecuador), not the Inka.[15] No artifact of this type has ever been conclusively assigned to the Inka.[16] These devices were perhaps architectural models for large constructions[17] or military purposes,[18] or were used in gaming.[19]

A second type of *yupana* is the checkerboard (Fig. 36.3, right) drawn by Felipe Guamán Poma de Ayala,[20] an Inka nobleman whose illustrated chronicle is known for its comprehensive and sympathetic insight into Inka society. However, there was disagreement as to whether the drawing represented a *yupana*, at least initially.[21] The diagram appears next to an Inka accountant holding a *khipu*, context positioning it as the calculating device. There is considerable disagreement over how it might have been used. Most scholars have assumed it shows positional numbers. Some interpret the black circles as counters and the white ones as empty holes, while others give both colors (and also the columns and rows) different numerical values and occasionally non-decimal bases, usually adding but sometimes subtracting them or rotating the board.[22] A unique "additive" interpretation by Italian researcher Cinzia Florio rearranges the counters to consolidate like colors and assumes one white circle was filled in by mistake; the result is a multiplication (5 × 32)[23] that resembles a method used in ancient Egypt.[24]

The stone floors of some Inka storehouses are pitted with regularly spaced holes, suggesting a possible third *yupana* type. If these floors were *yupana*, they might have been used in public accounting—for example, collecting and distributing taxes

[15] De Velasco (1789), as republished in (1841).

[16] Lau (2016).

[17] De Betánzos (1551).

[18] De Gamboa (1572); also see Pietschmann (1906).

[19] Bennett (1949); de Voogt (1995); Gonzalez Suarez (1892); Means (1931); Nordenskiöld (1918); Protzen (2011); Smith (1977).

[20] Guamán Poma (1615), p. 360.

[21] Ascher (1986); Ascher and Ascher (1981).

[22] Burn Glynn (1981); Calderón (1966); de Pasquale (2001); Kak (2014); Leonard and Shakiban (2010); Mendizábal Losack (1971); Moscovich (2007); Radicati di Primeglio (1979); Rojas-Gamarra and Stepanova (2015); Saldívar Olazo et al. (2019); Wassén (1940).

[23] Florio (2009).

[24] Damerow (2010); Reimer (2014); also see Florio and Overmann (2025).

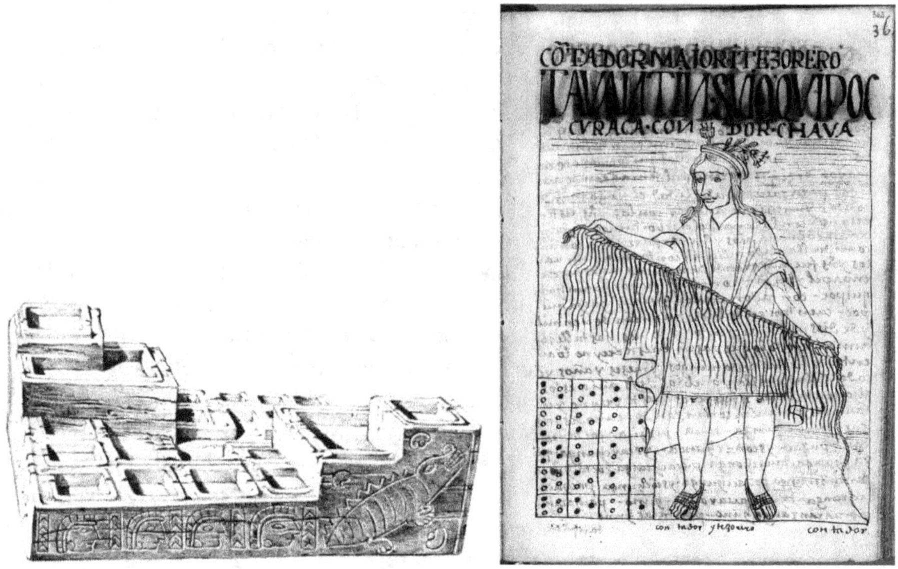

Fig. 36.3 *Yupana* designs. (Left) Drawing of a multilevel partitioned artifact found in Cuenca, Ecuador. Image from Gonzalez Suarez (1892, Pl. III) in the public domain. (Right) Drawing of an Inka accountant holding a *khipu* with the "checkerboard" *yupana* by his feet. Image from Guamán Poma (1615, p. 360) in the public domain

before crowds of witnesses.[25] However, their layouts mimic the storehouse complexes of which they are a part, rather than the linear layout more typical of abacuses, suggesting they may have served as rearrangeable keys to storehouse contents. Further, any grains or pebbles used as counters would be difficult to see and move in calculating because of the size, depth, and spacing of the holes. Future research will hopefully settle the debate over the form and function of the Inka *yupana*.

References

Ascher, M. (1986). Mathematical ideas of the Incas. In M. P. Closs (Ed.), *Native American mathematics* (pp. 261–289). University of Texas Press.

Ascher, M., & Ascher, R. (1981). *Code of the quipu: A study in media, mathematics, and culture*. University of Michigan Press.

Barraza Lescano, S., Areche Espinola, R., & Marcone Flores, G. (2022). By stones and by knots: The counting and recording of chili peppers stored during the Inca occupation of the Guarco administrative center of Huacones-Vilcahuasi, Lower Canete Valley, Peru. *Andean Past, 13*(1), 221–264.

[25] Barraza Lescano et al. (2022); Eeckhout (2012).

Bennett, W. C. (1949). Mnemonic and recording devices. In J. H. Steward (Ed.), *Handbook of South American Indians. Vol. 5: Comparative ethnology of the South American Indians* (pp. 611–619). U.S. Government.

Brokaw, G. (2010). *A history of the khipu*. Cambridge University Press.

Brotherston, G. (1995). Indigenous literatures and cultures in twentieth-century Latin America. In L. Bethell (Ed.), *The Cambridge history of Latin America. Vol. X: Latin America since 1930: Ideas, culture, and society* (pp. 287–305). Cambridge University Press.

Burn Glynn, W. (1981). La tabla de cálculo de los Incas [The calculation table of the Incas]. *Boletin de Lima, 11*(3), 1–15.

Calderón, H. M. (1966). *La ciencia matematica de los Mayas [The mathematical science of the Mayas]*. Editorial Orión.

Chrisomalis, S. (2010). *Numerical notation: A comparative history*. Cambridge University Press.

Damerow, P. (2010). *Abstraction and representation: Essays on the cultural evolution of thinking*. Kluwer Academic.

De Acosta, J. (1590). *Historia natural moral de las Indias, en que se tratan las cosas notables del cielo, y elementos, mentales, plantas, y animales dellas: y los ritos, y ceremonias, leyes, y gouierno, y guerras de los Indios [Natural and moral history of the Indies, in which the notable things of heaven, and the elements, minds, plants, and animals of them are treated: and the rites, and ceremonies, laws, and government, and wars of the Indians]*. Juan de Leon.

De Betánzos, J. (1551). *Suma y narracion de los Incas, que los indios llamaron capaccuna, que fueron señores de la ciudad del Cuzco y de todo lo á ella subject [Summary and narration of the Incas, whom the Indians called Capaccuna, who were lords of the city of Cuzco and of everything subject to it]*. Márcos Jiménez de la Espada.

De Gamboa, P. S. (1572). *Segunda parte de la Historia general llamada Índica, la cual por mandado del excelentísimo señor Don Francisco de Toledo, virrey, gobernador y capitán general de los reinos del Pirú y mayordomo de la casa real de Castilla, compuso el capitán Pedro Sarmien [Second part of the general history called Índica, which by order of the most excellent lord Don Francisco de Toledo, viceroy, governor and captain general of the kingdoms of Peru and steward of the royal house of Castile, was composed by Captain Pedro Sarmien]*. In R. Pietschmann (Ed.), *Geschichte des Inkareiches von Pedro Sarmiento de Gamboa (1906). Abhandlungen des Königlichen Gesellschaft der Wissenschaften zu Göttingen. Philogisch-Historische Klasse [History of the Inca Empire by Pedro Sarmiento de Gamboa (1906). Proceedings of the Royal Society of Sciences in Göttingen. Philological-Historical Class]* (Neue Folge, Band VL, pp. 1–138). Weidmannsche Buchandlung.

De la Vega, G. (1966). *Royal commentaries of the Incas, and general history of Peru. Part one* (H. V. Livermore, Trans.). University of Texas Press.

De Pasquale, N. (2001). Il volo del condor [The flight of the condor]. *Rivista dell'Ordine Degli Ingegneri della Provincia di Pescara, 0*, 1–19. http://www.quipus.it/DePasquale_Volo_del_Condor.pdf

De Velasco, G. (1789). *Viaggi, relazioni e memorie relative al regno di Quito [Travels, reports and memories relating to the kingdom of Quito]*.

De Velasco, J. (1841). *Historia del reino de Quito en la America meridional [History of the Kingdom of Quito in South America]* (Tome II, y Parte II). Juan Campuzano.

De Voogt, A. J. (1995). A classification of board games. In A. J. de Voogt (Ed.), *New approaches to board games research: Asian origins and future perspectives* (pp. 9–15). International Institute for Asian Studies.

Eeckhout, P. (2012). Inca storage and accounting facilities at Pachacamac. *Andean Past, 10*, 213–239.

Florio, C. (2009). Encuentros y desencuentros nella individuazione di una relazione matematica nella yupana in Guaman Poma de Ayala [Encounters and despairs in the identification of a mathematical relationship in the yupana of Guaman Poma de Ayala]. In E. Guagliano (Ed.), *Incontri e disincontri tra Europa e America, Atti del XXX Convegno Internazionale di Americanistica Salerno, 14–15 maggio e 10–12 dicembre 2008 [Meetings and discussions between Europe and America, Proceedings of the XXX International Conference of American Studies Salerno, 14–15 May and 10–12 December 2008]* (pp. 151–186). Oèdipus Editore.

Florio, C., & Overmann, K. A. (2025). Inka numbers, *khipu*, and *yupana*: A reanalysis. *Journal of Cognition and Culture, 25*(1–2), 128–158.

Gonçalez Holguin, D. (1607). *Gramatica y arte dveva de la lengva general de todo el Peru, llamada lengua Qquichua, o lengua del Inca [Grammar and art of the general language of all Peru, called the Quichua language, or the language of the Inca].* Francisco del Canto.

Gonzalez Suarez, F. (1892). Lamina tercera [Third plate]. In *Tiempos antiguos ó El Ecuador antes de la Conquista, Segunda parte, texto; atlas arqueologico, historia general de la Republica del Ecuador [Ancient times or Ecuador before the Conquest, Part two, text; archaeological atlas, general history of the Republic of Ecuador]* (pp. 67–75). Imprenta del Clero.

Guamán Poma, F. d. A. (1615). *El primer nueva corónica y buen Gobierno [The first new chronicle and good government].* Museum Tusculanum Press.

Hyland, S. (2020). Indicators and the decipherment of genre on Andean khipus. *Anthropological Linguistics, 62*(2), 137–158.

Hyland, S., Bennison, S., & Hyland, W. P. (2021). Khipus, khipu boards, and sacred texts: Toward a philology of Andean knotted cords. *Latin American Research Review, 56*(2), 400–416.

Kak, S. (2014). Guaman Poma's yupana and Inca astronomy. Preprint. In *arXiv* (Vol. 1401.7637). https://arxiv.org/ftp/arxiv/papers/1401/1401.7637.pdf

Lau, G. F. (2016). *An archaeology of Ancash: Stones, ruins and communities in Andean Peru.* Routledge.

Leonard, M., & Shakiban, C. (2010). The Incan abacus: A curious counting device. *Journal of Mathematics and Culture, 5*(2), 81–106.

Locke, L. L. (1912). The ancient quipu, a Peruvian knot record. *American Anthropologist, 14*(2), 325–332.

Martin, C. E., & Wasserman, M. (2010). Writing and record-keeping in the ancient Americas: The quipus of the Andes. In *Latin America and its people. Vol. 1: To 1830* (2nd ed., pp. 1–13). Prentice Hall.

Means, P. A. (1931). *Ancient civilizations of the Andes.* Charles Scribner's Sons.

Medrano, M., & Urton, G. (2018). Toward the decipherment of a set of mid-colonial khipus from the Santa Valley, Coastal Peru. *Ethnohistory, 65*(1), 1–23.

Mendizábal Losack, E. (1971). *Estructura y función en la cultura andina. Inka phase [Structure and function in Andean culture. Inka phase].* Unpublished doctoral dissertation, Universidad Nácional Mayor de San Marco.

Moscovich, V. (2007). Yupana, tabla de contar inca: Estructura interna [Yupana, Inca counting board: Internal structure]. *Revista Andina, 44*, 71–116.

Nordenskiöld, E. (1918). Spieltische aus Peru und Ecuador [Gaming tables from Peru and Ecuador]. *Zeitschrift für Ethnologie, 50*(2/3), 166–171.

Pietschmann, R. (1906). Geschichte des Inkareiches von Pedro Sarmiento de Gamboa (1572) [History of the Inca Empire by Pedro Sarmiento de Gamboa (1572)]. In *Abhandlungen des Königlichen Gesellschaft der Wissenschaften zu Göttingen. Philogisch-Historische Klasse [Proceedings of the Royal Society of Sciences in Göttingen. Philological-Historical Class]* (Neue Folge, Band VL, pp. III–CXVIII, 1–138). Weidmannsche Buchandlung.

Protzen, J.-P. (2011). Antiguas maquetas arquitectónicas: Qué nos pueden decir? [Ancient architectural models: What can they tell us?]. In C. Pardo (Ed.), *Modelando el mundo: Imágenes de la arquitectura precolombina [Modeling the world: Images of pre-Columbian architecture]* (pp. 84–109). Asociación Museo de Arte de Lima.

Radicati di Primeglio, C. (1979). *El sistema contable de los Incas: Yupana y quipu [The Inca accounting system: Yupana and quipu].* Libreria Studium.

Reimer, D. (2014). *Count like an Egyptian: A hands-on introduction to ancient mathematics.* Princeton University Press.

Rojas-Gamarra, M., & Stepanova, M. (2015). Sistema de numeración Inka en la yupana y el khipu [Inka numbering system in the yupana and the khipu]. *Revista Latinoamericana de Etnomatemática: Perspectivas Socioculturales de la Educación Matemática, 8*(3), 46–68.

Saldívar Olazo, C. G., Saldívar Olazo, A. J., & Goycochea Olazo, D. (2019). Tawa pukllay – la aritmética Inca de reconocimiento de formas y movimientos operable en paralelo y que no requiere cálculos numéricos mentales [The Inca arithmetic of shape and movement recognition operable in parallel and not requiring mental numerical calculations]. *Acta Latinoamericana de Matemática Educativa, 32*(1), 354–363.

Smith, J. W. (1977). Recuay gaming boards: A preliminary study. *Indiana, 4*, 111–137.

Urton, G. (2010). *The social life of numbers: A Quechua ontology of numbers and philosophy of arithmetic*. University of Texas Press.

Urton, G. (2017). *Inka history in knots: Reading khipus as primary sources*. University of Texas Press.

Urton, G. (2018). Quipus and yupanas as imperial registers: Reckoning and recording in Tahuantinsuya. In S. Alconini & R. A. Covey (Eds.), *The Oxford handbook of the Incas* (pp. 601–622). Oxford University Press.

Wassén, H. (1940). El antiguo ábaco peruano según el manuscrito de Guaman Poma [The ancient Peruvian abacus according to the Guaman Poma manuscript]. *Ethnologiska Studier, 11*, 1–30.

Williams, R. (1643). *A key into the language of America: Or, an help to the language of the natives in that part of America, called New-England*. Gregory Dexter.

Chapter 37
Amazonian Number Systems: Desana, Mundurukú, and Pirahã

Abstract The Amazon is home to Indigenous societies that are small and relatively isolated, conditions associated with *restricted* numbers, systems that count no higher than *twenty*. Desana have a quinary system that counts to *twenty* based on the fingers and toes. The Desana numbers *eleven* through *nineteen* are unrecorded, suggesting they are "gap numbers" that have not yet been explicitly named. Desana invented a system of numerical signs influenced by Western numbers and the idea of writing. Mundurukú count to *about four* using "distributed exemplars," common environmental objects associated with quantities that are small (perceptually appreciable) and reliable (for example, human arms always occur in pairs). Pirahã may not count at all: they have been described as lacking both words and gestures for numbers. The Pirahã language has also been described as lacking *recursion*, the ability to embed words and phrases that is used in constructing higher numbers (for example, as *twenty-three* is constructed from *two*, *ten*, and *three*). The extent of Pirahã anumeracy remains contested. Restricted numbers are providing new mathematical insights, as for example, the nature of infinity.

Keywords Restricted number systems · Distributed exemplars · Gap numbers · Desana · Mundurukú · Pirahã

Well to the east of Inka territory and separated by impassible terrain lies the Amazonian region that is home to many Indigenous societies. These societies are largely untouched by external cultures, know little of external societies, and have low immunity to common diseases.[1] Such societies have few or possibly no numbers[2] because the conditions motivating the need for numbers are associated with large societies and intergroup contact.[3] Like Australia, South America is one of the

[1] Da Silva and Bergo Leugi (2022).

[2] Divale (1999); Epps (2006); Epps et al. (2012); Hammarström (2015); Heine (1997); Stampe (1976); Winter (1999).

[3] Overmann (2023).

K. A. Overmann, *Cultural Number Systems*, Interdisciplinary Contributions to Archaeology, https://doi.org/10.1007/978-3-031-83383-0_37

Table 37.1 Desana numbers (compiled from Miller, 1999; de Silva, 2012)

Nr.	Desana numbers	Meaning
1	*yuhu*	*one* (1)
2	*pẽ-*	*two* (2)
3	*ɨ̃ɾ̃ẽ-*	*three* (3)
4	*ahpikɨ̃ɾ̃ã*	*two pairs* (2 + 2)
5	*yuhuru mõhõtõ*	*one hand* (5)
6	*yuhuru mõhõtõ yuhuru nĩã̃ɾ̃ã*	*one hand, one from other hand* (5 + 1)
7	*yuhuru mõhõtõ pẽɾ̃ʉ̃ nĩã̃ɾ̃ã*	*one hand, two from other hand* (5 + 2)
8	*yuhuru mõhõtõ ɨ̃ɾ̃ẽɾ̃ʉ̃ nĩã̃ɾ̃ã*	*one hand, three from other hand* (5 + 3)
9	*yuhuru mõhõtõ ahpikɨ̃ɾ̃ã nĩã̃ɾ̃ã*	*one hand, four from other hand* (5 + 4)
10	*pẽmõhõtõ*	*two hands* (5 + 5)
20	*guburi peʔre-ri bãhã-*	*one man* (20)

most recently colonized places on the planet, so the Amazonian number systems are thought to be in the process of emerging. Here we look at three.

The Desana of the Uaupés River region[4] *count to* twenty.[5] These numbers (Table 37.1) are classified as a quinary digit-tally system: *quinary* because counting proceeds by groups of *five* (but not yet by groups of *twenty*), and *digit-tally* because the human digits are used as a tally. The term *maha*, meaning *a lot*, is used to express the idea of values higher than *twenty*.[6] The names of most Desana numbers are analyzable as phrases related to the hand, with the highest number representing all the digits of a single person. Analyzability is characteristic of recent emergence, infrequent use, or both. Since the numbers *one* through *three* are short and unanalyzable, they are likely older than the other numbers. This reflects the typical order of emergence: numbers in the "subitizable" (perceptually appreciable) range of *one* to *about four* are always first because of their perceptibility; the first numbers to emerge beyond the subitizing range are usually *five* and *ten*, as scaffolded by the hands, and these in turn are followed by the "gap numbers" *four* and *six* through *nine*, as scaffolded by the intervening fingers.[7] Since the Desana numbers *eleven* through *nineteen* are unrecorded, they may be higher "gap numbers" that have yet to be explicitly named.

The Desana don't seem to use material tallies like notched sticks or knotted strings. However, they have invented a unique system of numerical signs[8] that leverages Western decimal numbers and the idea of writing.[9] The way these numerical signs are constructed not only blends aspects of the Indigenous quinary system with

[4] *Desana* (2024).

[5] Miller (1999); de Silva (2012).

[6] Miller (1999), p. 46.

[7] Overmann (2023).

[8] Fernandes and Fernandes (2006).

[9] Overmann et al. (2022).

Table 37.2 Mundurukú numbers (Rooryck et al., 2017)

Nr.	Mundurukú term	Meaning	Number	Syllables
1	*pûg*	"one"	*One*	One
2	*xep-xep*	"fat" (reduplicated)	*Two*	Two
3	*eba-pûg*	"your arm(s) and one"	*About three*	Three
4	*eba-dip-dip*	"your parent(s)" (reduplicated)	*About four*	Four

European decimal numbers, it also implies the absence of tallying, which accumulates identical elements placed in one-to-one correspondence with whatever is being counted.[10] Instead of being identical, each element in a Desana numerical sign differs from its fellows in its shape, size, and placement, creating unique combinations that must be memorized.[11] In addition, the elements are not necessarily accumulated in one-to-one correspondence with the items being counted, as for example, a three-pointed sign expresses the idea of *four men*.[12]

The Mundurukú of the Tapajós River Valley[13] *count to* about four.[14] These numbers (Table 37.2) involve "distributed exemplars," common environmental objects associated with subitizable quantities.[15] Exemplars of *two* include the eyes (Hup, another Amazonian group) and ears (Chinese);[16] the Mundurukú use the human arms in their term for *about three*. The Mundurukú numbers for *two*, *about three*, and *about four* are compounds that build larger quantities by combining smaller ones. The highest numbers are approximate or "fuzzy": *about three* and *about four*. This suggests a perceptual effect that occurs as quantities approach the upper limit of the subitizing range: they become indistinguishable, so they are expressed simply as "many." Above *about four*, the Mundurukú use the idea of a handful.[17] This suggests the transition to using the fingers as a counting device, though initially, the Mundurukú may be involving the hand more for what it can grasp than for the quantity of its fingers. Using the fingers to count makes numbers discrete, as it's difficult for a number represented by a specific finger to remain fuzzy. Most intriguingly, the Mundurukú number terms have the corresponding number of syllables,[18] yielding a multimodal expression of numbers: the audible quantity of syllables and the visible quantity of objects leveraged as or compared to exemplars. This, in turn, suggests that verbal forms—even when their syllables are numerically informative—don't influence numbers toward conceptual discreteness, since *about three* and *about four* remain fuzzy.

---

[10] Overmann et al. (2022); also see discussion of tokens in Schlaudt (2020).

[11] Fernandes and Fernandes (2006); also see Overmann et al. (2022).

[12] Fernandes and Fernandes (2006); also see Overmann et al. (2022).

[13] *Munduruku* (2024).

[14] Pica and Lecomte (2008); Rooryck et al. (2017).

[15] Overmann (2023).

[16] Epps et al. (2012); Gow (1884).

[17] Pica and Lecomte (2008).

[18] Rooryck et al. (2017).

The Pirahã inhabiting the Madeira river region[19] *may not count at all.*[20] This absence of numbers is contested, as the Pirahã may have terms for *one* and perhaps even *two.*[21] As described by the linguist Daniel Everett, who lived with the Pirahã long enough to become expert in their language and culture,

> There is one word, "hói," which does come close to the numeral 1. But it can also mean "small" or describe a relatively small amount—like two small fish as opposed to one big fish, for example. And they don't even appear to count without language, on their fingers for example, in order to determine how many pieces of meat they have to grill for the villagers, how many days of meat they have left from the anteaters they've hunted or how much they demand from Brazilian traders for their six baskets of Brazil nuts.
> Von Bredow (2006)

Besides lacking numbers, the Pirahã language is also said to lack *recursion*, the ability to embed words, clauses, or phrases beside or within similar structures.[22] Recursion is thought to be the specific cognitive mechanism that generates "an infinite range of expressions from a finite set of elements" in language, and as such has been proclaimed the "only uniquely human component of the faculty of language."[23] However, this position is seriously challenged by the fact that Pirahã gets along perfectly well without recursion[24]—as can languages like English, as typified by the short, snappy sentences found in the work of authors like Ernest Hemingway.[25] Researchers have tried to associate the lack of recursion in Pirahã with the lack of numbers.[26] However, as the Mundurukú and Desana examples showed, the initial number-words always stop at the upper limit of the subitizing range, with the first non-subitizable numbers emerging later as scaffolded by the hands; if recursion were involved, the inevitable pause after the subitizable numbers becomes very difficult to explain, as a recursive mechanism wouldn't stop. Further, if recursion isn't needed for the initial number-words, the lack of recursion wouldn't be a barrier to their emergence in Pirahã. And finally, Pirahã men (but apparently not Pirahã women because of different cultural expectations) also learn Portuguese, a language with recursion,[27] and this means that whatever the reason for the lack of recursion in Pirahã, it certainly isn't a cognitive difference.

There is another unique aspect to Pirahã culture: they don't keep track of time.[28] Early timekeeping involves counting the recurrence of lunar and/or solar units like

---

[19] *Pirahã* (2024).

[20] Everett (2005, 2007, 2013); Frank et al. (2008).

[21] Frank et al. (2008); Gordon (2004).

[22] Everett (2005); Sakel and Stapert (2010).

[23] Hauser et al. (2002), p. 1569.

[24] Liberman (2006).

[25] Liberman (2005).

[26] Everett (2005); Frank et al. (2008).

[27] Sakel (2012).

[28] Von Bredow (2006).

days, months, and years, and this is an activity for which numbers are essential.[29] Simply, when there aren't any numbers, nothing gets counted, including things like time and human age.[30]

Restricted number systems are providing new insights in mathematics. An example starts with paradoxes of infinity like the one noted in 1638 by Galileo, who observed that all numbers can be squared, but not all numbers are squares. As sets of numbers, the first seems larger than the second, and yet both are infinite in size—a paradox.[31] Galileo concluded the idea that sets could have different sizes could be applied only to finite sets, not infinite ones. In languages like Pirahã and Mundurukú, if *one* or *two* were added to *many*, the result would be *many*; this result is "not wrong," "just inaccurate," and the inaccuracy will disappear once the *many* has been specified as discrete numbers.[32] This insight, in turn, suggests that infinity is a parallel case, and that its inaccuracy will similarly disappear once it has been specified to reveal the differences it contains.[33]

References

Clark, M. (2012). Galileo's paradox. In *Paradoxes from A to Z* (3rd ed., pp. 74–78). Routledge.

Da Silva, F. B., & Bergo Leugi, G. (2022). Behavioral community psychology in the Amazon rainforest: Suggestions for when behavior analysts meet alterity. *Behavior and Social Issues, 31*(1), 234–251.

Desana. (2024). *Povos Indígenas no Brasil [Indigenous peoples in Brazil]*. https://pib.socioambiental.org/en/Povo:Desana

Divale, W. (1999). Climatic instability, food storage, and the development of numerical counting: A cross-cultural study. *Cross-Cultural Research, 33*(4), 341–368.

Epps, P. (2006). Growing a numeral system: The historical development of numerals in an Amazonian language family. *Diachronica, 23*(2), 259–288.

Epps, P., Bowern, C., Hansen, C. A., Hill, J. H., & Zentz, J. (2012). On numeral complexity in hunter-gatherer languages. *Linguistic Typology, 16*(1), 41–109.

Everett, D. L. (2005). Cultural constraints on grammar and cognition in Pirahã: Another look at the design features of human language. *Current Anthropology, 46*(4), 621–646.

Everett, D. L. (2007). *Cultural constraints on grammar in Pirahã: A reply to Nevins, Pesetsky, and Rodrigues*. Unpublished manuscript, https://ling.auf.net/lingbuzz/000427

Everett, D. L. (2013). Recursion and human thought: Why the Pirahã don't have numbers. In J. Brockman (Ed.), *Thinking: The new science of decision-making, problem-solving, and prediction* (pp. 269–291). HarperCollins.

Fernandes, A. C., & Fernandes, D. M. (2006). *Bueri kãdiri marĩriye: Os ensinamentos que não se esquecem [The teachings that are not forgotten]*. UNIRT/FOIRN.

Frank, M. C., Everett, D. L., Fedorenko, E., & Gibson, E. (2008). Number as a cognitive technology: Evidence from Pirahã language and cognition. *Cognition, 108*(3), 819–824.

[29] Overmann (2013, 2024).

[30] Overmann (2013).

[31] Clark (2012).

[32] Sergeyev (2022), p. 9.

[33] Horák et al. (2023); Sergeyev (2022).

Gordon, P. (2004). Numerical cognition without words: Evidence from Amazonia. *Science, 306*(5695), 496–499.

Gow, J. (1884). *A short history of Greek mathematics*. Cambridge University Press.

Hammarström, H. (2015). *Restricted numeral systems and the hunter-gatherer connection*. Max Planck Institute for Psycholinguistics. https://www.eva.mpg.de/fileadmin/content_files/linguistics/conferences/2015-speaking-of-Khoisan/P8b_Linguistics_Hammarstroem_Numeral_systems.pdf

Hauser, M. D., Chomsky, N., & Fitch, W. T. (2002). The faculty of language: What is it, who has it, and how did it evolve? *Science, 298*(5598), 1569–1579.

Heine, B. (1997). *Cognitive foundations of grammar*. Oxford University Press.

Horák, M., Uhrin, M., & Amaral, A. P. (2023). Status of the current scientific knowledge on Pirahã: What is known and what could be studied in future? *Anthropologia Integra, 14*(2), 25–33.

Liberman, M. (2005). *Homo hemingwayensis. Language Logs*. University of Pennsylvania. http://itre.cis.upenn.edu/~myl/languagelog/archives/001792.html

Liberman, M. (2006). *Parataxis in Pirahã. Language Logs*. University of Pennsylvania. http://itre.cis.upenn.edu/~myl/languagelog/archives/003162.html

Miller, M. (1999). *Desano grammar*. Summer Institute of Linguistics and the University of Texas at Arlington.

Munduruku. (2024). *Povos Indígenas no Brasil [Indigenous peoples in Brazil]*. https://pib.socioambiental.org/en/Povo:Munduruku

Overmann, K. A. (2013). Material scaffolds in numbers and time. *Cambridge Archaeological Journal, 23*(1), 19–39.

Overmann, K. A. (2023). *The materiality of numbers: Emergence and elaboration from prehistory to present*. Cambridge University Press.

Overmann, K. A. (2024). The beginning of time. *Cambridge Archaeological Journal, 34*(4), 693–709.

Overmann, K. A., Chacon, T., & Payne, A. (2022). Desana numerical symbols: An indigenous creation narrated by Diakuru and Kisibi. *Written Language and Literacy, 25*(2), 133–158.

Pica, P., & Lecomte, A. (2008). Theoretical implications of the study of numbers and numerals in Mundurucu. *Philosophical Psychology, 21*(4), 507–522.

Pirahã. (2024). *Povos Indígenas no Brasil [Indigenous peoples in Brazil]*. https://pib.socioambiental.org/en/Povo:Pirahã

Rooryck, J., Saw, J., Tonda, A., & Pica, P. (2017). *Mundurucú number words as a window on short-term memory*. The Royal Society of London. https://halshs.archives-ouvertes.fr/halshs-01497577

Sakel, J. (2012). Acquiring complexity: The Portuguese of some Pirahã men. *Linguistic Discovery, 10*(1), 75–88.

Sakel, J., & Stapert, E. (2010). Pirahã–In need of recursive syntax? In H. Van der Hulst (Ed.), *Recursion and human language* (pp. 3–16). Walter de Gruyter.

Schlaudt, O. (2020). Type and token in the prehistoric origins of numbers. *Cambridge Archaeological Journal, 30*(4), 629–646.

Sergeyev, Y. D. (2022). Some paradoxes of infinity revisited. *Mediterranean Journal of Mathematics, 19*, 1–28.

Silva, W. d. L. (2012). *A descriptive grammar of Desano*. University of Utah.

Stampe, D. (1976). Cardinal number systems. In S. S. Mufwene, C. A. Walker, & S. B. Steever (Eds.), *Papers from the twelfth regional meeting, Chicago Linguistic Society, April 23–25, 1976* (Vol. 12, pp. 594–609). Chicago Linguistic Society.

Von Bredow, R. (2006). *Living without numbers or time*. SPIEGEL International. https://www.spiegel.de/international/spiegel/brazil-s-piraha-tribe-living-without-numbers-or-time-a-414291.html

Winter, W. (1999). When numeral systems are expanded. In J. Gvozdanović (Ed.), *Numeral types and changes worldwide* (pp. 43–53). Mouton de Gruyter.

Chapter 38
Conclusion

Abstract Historically, few researchers have given much thought to the cognitive role of the objects used to represent and manipulate numbers. We aren't used to thinking of material devices as being a meaningful part of our cognition, and as a result, we are literally blind to them. This makes it difficult to recognize that the material forms used in numbers influence how and what we think about them. Nonetheless, variability in cultural number systems is a matter of *whether* material forms are used, *which ones* are used, and *how* they are used. Material forms like fingers, tallies, tokens, and written numerals are the key to cross-cultural numerical variability because, as a single species, everything else that contributes to numbers is the same. Despite our profound neurological, physiological, and behavioral similarity, not all human societies count, and of those that do, not all count in the same way. This variability suggests that humans don't come pre-equipped with internal instructions for realizing and then structuring and organizing their numbers. The conclusion ends with a list of questions that researchers in different fields might ask in response to the cross-cultural variability found in number systems.

Keywords Cross-cultural variability · Number systems · Material forms · Piaget · Plato

If there's an overarching theme to this book, it's that variability in cultural number systems is a matter of *whether* material forms are used, *which ones* are used, and *how* they are used. Fingers, tallies, tokens, and written numerals are the key to cross-cultural numerical variability because everything else in numbers is the same. As a single human species, we have essentially the same neurological and physiological makeup and behavioral dispositions. These shared resources give all human societies the same starting point for numbers: the same ability to appreciate quantity and the same impulse to count on the hands.

Despite this profound similarity, not all human societies count, and of those that do, not all count in the same way. This variability suggests we don't come pre-equipped with internal instructions for realizing and then structuring and organizing

K. A. Overmann, *Cultural Number Systems*, Interdisciplinary Contributions to Archaeology, https://doi.org/10.1007/978-3-031-83383-0_38

our numbers. If the brain doesn't provide invariant formatting for numbers,[1] it explains why human societies are able to make so many different decisions about how they should use their visually salient hands as a device for counting and what other material forms they could use once the fingers become insufficient in their capacity and persistence.[2] The result is a glorious cacophony of numerical differences, systematized by the material forms used and harmonized by the fact that the outcomes, both material and conceptual, remain mutually intelligible in a way languages never do.

In one sense, Plato was correct that numbers are discoverable because any society that puts two things together with two other things will eventually conclude, at the point in its history where such things become useful, that $2 + 2 = 4$. The relations are in the quantities and the acts of moving their representations into new patterns—at least for any species capable of making the movements and apprehending the relations and patterns created thereby. On the other hand, Plato's idea that numbers are pre-existing, full-fledged entities that we somehow come into contact with doesn't seem right. Each society with numbers explicates and elaborates them by manipulating whatever material forms it has chosen to represent them, a slow and laborious process. In the case of $2 + 2 = 4$, it took thousands of years—likely tens of thousands—to become able to express the idea as a succinct equation with symbolic numerals and semasiographic operating signs.[3] Considered across cultures, numerical outcomes are not just highly variable, they also require millennia of effort and refinement. If full-fledged numbers were "out there" in the way Plato thought they were, the process through which they are achieved is inexplicable, as is the cross-cultural variability it yields.

Piaget was also right to claim that the human brain is critical to numbers. Other species neither count on their fingers nor systematically rearrange material quantities into new configurations in order to understand them better. The fact that our species does these things is not just a significant difference in behavior, it's a substantial cognitive difference as well. Material representations of quantity and their manipulations provide opportunities for relations and patterns to be recognized and then named if they are useful and recur with sufficient frequency, and these are things the human brain does very well, on the conditions that it receives adequate stimulus and the society finds the ideas useful.

Nonetheless, Piaget's idea that the brain provides *invariant* numerical formatting requires some nuance at the very least, given that cross-cultural numerical outcomes cannot justly be characterized as such. Moreover, the notion that numbers must either be purely external (Plato's realism[4]) or purely internal (Brouwer's

[1] Overmann (2024b).

[2] Overmann (2016).

[3] Schulte (2015).

[4] Linnebo (2018); Maddy (1990).

intuitionism[5] and the perspective of Piaget[6] and the modern cognitive sciences[7]) is almost certainly wrong. In involving material forms and psychological processes as equal partners—albeit ones performing different, complementary roles—numbers appear to be both, with the hands acting to bridge the external and internal domains.[8] This is the perspective and model adopted for the analyses here, the extended mind perspective of Material Engagement Theory.[9]

Ever since Plato's day, thinkers across multiple disciplines have been trying to explain what numbers are as concepts and how societies get them. But before Malafouris observed that material forms make quantity tangible and thus tractable to manipulation,[10] no one seems to have given much thought to the cognitive role of the objects used to represent and manipulate numbers. Recognizing that the material forms used in numbers influence how and what we think about them is difficult.[11] We aren't used to thinking of material devices as being a meaningful part of our cognition, and as a result, we are literally blind to them. Even in accounts of numerical origins that prominently list material devices like tallies and tokens,[12] they are still not seen as anything more than passive repositories for mentally originated concepts, and in cognitivist accounts, they are hardly mentioned at all.[13] I hope the analyses presented here have clearly and convincingly drawn the connection between the concepts of numbers and the material forms used to represent and manipulate them.

Speaking of material forms, it's worth pointing out that—for Western culture—symbolic numerals are widely seen as having a role in how numerical concepts are acquired and understood.[14] My argument about what numbers are and how they are realized simply extends this role to the other material forms used—the hand and objects like tallies and tokens. Another thing I hope is clear from reading this account is that numbers are so much broader a category than written symbols and spoken words, even if those are the primary modalities through which numbers are experienced in Western culture on a day-to-day basis.

What questions should we be asking in response to the variability in cultural number systems? In mathematics, we could be investigating what we might learn from the different cultural systems. For example, highly restricted numbers, despite not counting beyond the perceptually appreciable range of *about three or four*,

---

[5] Brouwer (1981).

[6] Piaget (1928, 1950, 1952).

[7] Nieder (2017).

[8] Gallagher (2013).

[9] Malafouris (2013).

[10] Malafouris (2010); also see discussion in Coolidge and Overmann (2012).

[11] Overmann (2021).

[12] E.g., Ifrah (1981, 1985, 2000).

[13] E.g., Lakoff and Núñez (2000).

[14] Schlimm (2018).

provide insights applicable to complex concepts like infinity.[15] Similar insights are surely possible. Our greatest opportunities for acquiring new insights into numbers is likely to come from cultural systems with the fewest of them, for they are the ones with the greatest variability and fewest preconceptions, and thus, the best potential for demonstrating the range of the human capacity for numbers.

In linguistics and psychology, we could be looking at the ways in which our capacity for conceiving numbers and expressing them in language absorb characteristics from the devices used as influenced by the manner chosen for their use. Some of us count to *ten* on our hands, while others count to *four*, *five*, *six*, *eight*, *twelve*, or *fourteen*, and these amounts presumably reflect local choices made in the absence of an internal (brain-bound) source of structure. But how does the brain incorporate the chosen amount as a parameter? We don't know because the cognitive sciences aren't looking into it. This isn't just a limitation of methods, the question of how to investigate the role of material forms in numerical cognition, both as a component of contemporary psychological responses to stimuli and as a historical influence on conceptualization. It's also a limitation of theory, the decision to see material forms as integral to numerical cognition and investigate them as such, when the state-of-the-art is focused on the brain. Theory is further limited by seeing the involvement of the fingers as a mundane aspect of embodiment,[16] an assumption that admits the fingers but keeps numbers firmly sequestered inside the head. But congenitally blind people don't count on their fingers, not even idiosyncratically, and this suggests there's no internal (brain-bound) impetus for finger-counting, and without it, no internal means of accessing whatever important and foundational structuring the brain might provide to a number system.[17]

In ethnography, we could be documenting numerical devices and behaviors in much greater detail. Data collection is currently focused on numerical vocabularies, while other aspects of numbers are generally ignored. Speaking as someone who spends hours combing through the ethnographic literature for traces of numbers, I can attest to the fact that few things are more frustrating than finding only hints and gaps where I'd most like to find rich descriptions and high-resolution images of counting devices and behaviors. But we should also recognize why these data are critical: counting materials and methods are the mechanism for numerical emergence and elaboration, and this means we cannot understand numerical emergence and elaboration unless we understand the mechanism through which they occur.

For archaeology, we need to think even more deeply about the cognitive value of material culture. The theoretical paradigm I use, Material Engagement Theory,[18] is a viable framework for investigating the ancient mind through its material component—and for that matter, for investigating how our own minds work today and how they might work in the future. Material Engagement Theory is still relatively new,

[15] Sergeyev (2022).

[16] Johnson and Everett (2021).

[17] Overmann (2023); also see Marlair et al. (2024).

[18] Malafouris (2013).

and as such, only time will tell what insights the paradigm may ultimately yield. As a discipline, however, cognitive archaeology tends to focus on the remote ancestral past, where things like stone tools provide some insight into the gradual processes of change that ultimately produced humanity from its non-human antecedents. This temporal focus means the very recent past—especially the last 5000 years—is generally overlooked as evolutionarily uninteresting. But it is precisely because of its recency that this timeframe has the greatest potential to reveal *cultural* change, the ways in which our interactions with material culture have changed not just our tools but also our behaviors and brains. Moreover, the insights we gain by studying cognitive change in the recent past are applicable beyond the last 5000 years—to our present, to our future, and even to the millions of years of the past in which our bipedal ancestors were using tools before us.[19]

Since my formal study of numerical devices began with the Mesopotamian tokens, I want to mention some other archaeological matters that I see as shortfalls. As has been noted in the literature,[20] small clay objects have been under-collected, under-documented, and under-analyzed more often than not. Let's say hypothetically that these objects were being adequately collected and documented. The next problem is determining whether or not they were used as numerical counters. Here the problem is that there are no reliable criteria for making such determinations, and non-numerical uses are often plausible and likely.[21] Let's say for the sake of argument that we have criteria and can determine the objects were in fact numerical counters. Unfortunately, unless we also have associated impressions on bullae and tablets to tell us how they were grouped and ordered—important information that fixed impressions preserve but loose tokens don't—tokens cannot provide any numerical information beyond the simple fact they were used as counters (at which point we return to the difficulty of determining this in the first place).

These problems apply to more than the Mesopotamian tokens, as things like beads,[22] hand stencils,[23] and inscribed marks on bones[24] are also thought to represent numbers or counting, while similarly lacking in collection, documentation, and criteria.[25] Simply, we need new methods for determining whether objects had any numerical purpose, new ways of extracting numerical structure and organization from unaccompanied loose objects, and renewed commitment to methodical collection and documentation.

Prehistoric archaeology shares these challenges and more. First and foremost, we need to know what kinds of numbers we should be looking for. Prehistoric numbers, presumably, were newly emerged, so if we think they would have resembled

[19] Harmand et al. (2015); McPherron et al. (2010).

[20] E.g., Schmandt-Besserat (1992); also see discussion in Overmann (2019).

[21] Englund (1998); Friberg (1994); Zimansky (1993).

[22] D'Errico et al. (2005).

[23] Rouillon (2006).

[24] D'Errico et al. (2018).

[25] D'Errico (1998); Overmann (2023).

the newly emerged number systems of today—and there's no reason to assume otherwise—we should expect them to count to relatively low numbers, to be highly variable in how they counted, and to have counted in multiple ways.[26] These characteristics should influence what we look for and how we interpret what we find. We should also expect prehistoric peoples to have represented their numbers with objects that are ephemeral, made of materials that degrade quickly, or difficult to recognize as counting devices. Things like finger-counting, wood and plant fibers, and collections of pebbles are unlikely to leave much trace in the archaeological record in the first place, or to be recognized as counting materials if they do. This means we need to know why a society might start to use a less perishable material like bone to record its numbers. This is something we might gain insight into through the ethnographic data—especially if those data were more robust, a need noted earlier. We also currently lack an ethnographic standard that can serve as a basis for differentiating numerical marks from those made for the full range of non-numerical purposes (e.g., memorized lists, maps, music, decoration, and grip improvement), let alone identify the kind of number system a society once had by means of the marks it left behind.[27]

Finally, I am acutely aware of the shortfalls of the present study. Personal interest and assessments of uniqueness do not produce a systematic, comprehensive study. The gaps are many, and they are geographic, temporal, and cultural. I didn't, for example, include much material about prototypical decimal numbers, not just the Western numerals but also Indigenous ones from, for example, the middle of North America. Their omission wasn't because they don't exist, and it wasn't because I don't know about them. Rather, I didn't include them because I don't find prototypical decimal numbers very interesting.

As an archaeologist, my interest in numbers tends to be fairly narrow in scope—the materials and behaviors used in counting and how they inform numerical organization and structure. This means there are a lot of things about numbers I don't address, like their different social uses[28] or the cultural beliefs making some of them lucky.[29] Further, there are thousands of cultural number systems, and I only know a few hundred of them, so the potential for cross-cultural variability to be even greater than what I have been able to present here is quite likely. My personal wish list includes an online platform for collecting and freely sharing data on number systems—perhaps a Wikipedia-like model where anyone, subject to scholarly and community standards, could post whatever they know (and can adequately reference) about cultural number systems in all their different aspects.

My personal wish list also includes the hope that other researchers will take up these matters—not just the few questions I have posed here, but also the greater challenge of investigating cultural number systems while we still have them. These

[26] Overmann (2024a).

[27] Overmann (2024a).

[28] E.g., Urton (2010).

[29] E.g., Pellatt (2007).

systems are fast vanishing from the world through a variety of mechanisms, including cultural contact, political influence, population decrease, environmental displacement, language loss, and social convenience. Our window into this impressively variable human capacity is closing fast, and one day it will have shut completely. On the day that only Western numbers remain, I shall mourn.

References

Brouwer, L. E. J. (1981). *Brouwer's Cambridge lectures on intuitionism* (D. van Dalen, Ed.). Cambridge University Press.

Coolidge, F. L., & Overmann, K. A. (2012). Numerosity, abstraction, and the emergence of symbolic thinking. *Current Anthropology, 53*(2), 204–225.

D'Errico, F. (1998). Palaeolithic origins of artificial memory systems: An evolutionary perspective. In C. Renfrew & C. Scarre (Eds.), *Cognition and material culture: The archaeology of symbolic storage* (pp. 19–50). McDonald Institute for Archaeological Research.

D'Errico, F., Henshilwood, C. S., Vanhaeren, M., & Van Niekerk, K. L. (2005). *Nassarius kraussianus* shell beads from Blombos Cave: Evidence for symbolic behaviour in the Middle Stone Age. *Journal of Human Evolution, 48*(1), 3–24.

D'Errico, F., Doyon, L., Colagé, I., et al. (2018). From number sense to number symbols: An archaeological perspective. *Philosophical Transactions of the Royal Society of London. Series B, Biological Sciences, 373*(1740), 1–10.

Englund, R. K. (1998). Review: Denise Schmandt-Besserat, *How writing came about. Written Language and Literacy, 1*, 257–261.

Friberg, J. (1994). Preliterate counting and accounting in the Middle East: A constructively critical review of Schmandt-Besserat's *Before writing. Orientalistische Literaturzeitung, 89*(5–6), 477–489.

Gallagher, S. (2013). The enactive hand. In Z. Radman (Ed.), *The hand, an organ of the mind: What the manual tells the mental* (pp. 209–225). MIT Press.

Harmand, S., Lewis, J. E., Feibel, C. S., et al. (2015). 3.3-million-year-old stone tools from Lomekwi 3, West Turkana, Kenya. *Nature, 521*(7552), 310–315.

Ifrah, G. (1981). *The universal history of computing: From the abacus to the quantum computer* (R. Laffont, Trans.). Wiley.

Ifrah, G. (1985). *From one to zero: A universal history of numbers* (L. Bair, Trans.). Viking.

Ifrah, G. (2000). *The universal history of numbers: From prehistory to the invention of the computer* (D. Bellos, E. F. Harding, I. Monk, & S. Wood, Trans.). Wiley.

Johnson, M., & Everett, C. (2021). Embodied and extended numerical cognition. In A. Killin & S. Allen-Hermanson (Eds.), *Explorations in archaeology and philosophy* (pp. 125–148). Springer Nature.

Lakoff, G., & Núñez, R. E. (2000). *Where mathematics comes from: How the embodied mind brings mathematics into being*. Basic Books.

Linnebo, Ø. (2018). Platonism in the philosophy of mathematics. In E. N. Zalta (Ed.), *Stanford Encyclopedia of Philosophy (Spring 2021)*. Stanford University. https://plato.stanford.edu/archives/spr2018/entries/platonism-mathematics/

Maddy, P. (1990). *Realism in mathematics*. Clarendon Press.

Malafouris, L. (2010). Grasping the concept of number: How did the sapient mind move beyond approximation? In C. Renfrew & I. Morley (Eds.), *The archaeology of measurement: Comprehending heaven, earth and time in ancient societies* (pp. 35–42). Cambridge University Press.

Malafouris, L. (2013). *How things shape the mind: A theory of material engagement*. MIT Press.

Marlair, C., Guillon, A., Vynckier, M., & Crollen, V. (2024). Enhancing mathematics learning through finger-counting: A study investigating tactile strategies in 2 visually impaired cases. *Applied Neuropsychology: Child, 13*(3), 269–281.

McPherron, S. P., Alemseged, Z., Marean, C. W., et al. (2010). Evidence for stone-tool-assisted consumption of animal tissues before 3.39 million years ago at Dikika, Ethiopia. *Nature, 466*(7308), 857–860.

Nieder, A. (2017). Number faculty is rooted in our biological heritage. *Trends in Cognitive Sciences, 21*(6), 403–404.

Overmann, K. A. (2016). The role of materiality in numerical cognition. *Quaternary International, 405*, 42–51.

Overmann, K. A. (2019). *The material origin of numbers: Insights from the archaeology of the ancient Near East*. Gorgias Studies in the Ancient Near East 14. Gorgias Press.

Overmann, K. A. (2021). Numerical origins: The critical questions. *Journal of Cognition and Culture, 21*(5), 449–468.

Overmann, K. A. (2023). *The materiality of numbers: Emergence and elaboration from prehistory to present*. Cambridge University Press.

Overmann, K. A. (2024a). Prehistoric numeracy: Approaches, assumptions, and issues. In T. Wynn, K. A. Overmann, & F. L. Coolidge (Eds.), *The Oxford handbook of cognitive archaeology* (pp. 411–432). Oxford University Press.

Overmann, K. A. (2024b). Strange and wonderful: Numbers through a new (material) lens. *Cuneiform Digital Library Journal, 2*, 1–21.

Pellatt, V. (2007). *Numbers and numeracy in Chinese culture, language, and education: The social substratum of the development of mathematical thinking*. Edwin Mellen Press.

Piaget, J. (1928). Logique génétique et sociologie [Genetic logic and sociology]. *Revue Philosophique de la France et de l'Étranger, 105*, 167–205.

Piaget, J. (1950). *Introduction à l'epistemologie genetique: La pensée mathématique [Introduction to genetic epistemology: Mathematical thinking]*. Presses Universitaires de France.

Piaget, J. (1952). *The child's conception of number* (C. Gattegno & F. M. Hodgson, Trans.). Routledge.

Rouillon, A. (2006). Au Gravettien, dans la grotte Cosquer (Marseille, Bouches-du-Rhône), l'Homme a-t-il compté sur ses doigts? [In the Gravettian, in the Cosquer Cave (Marseille, Bouches-du-Rhône), did man count on his fingers?]. *L'Anthropologie, 110*(4), 500–509.

Schlimm, D. (2018). Numbers through numerals: The constitutive role of external representations. In S. Bangu (Ed.), *Naturalizing logico-mathematical knowledge: Approaches from psychology and cognitive science* (pp. 195–217). Routledge.

Schmandt-Besserat, D. (1992). *Before writing: From counting to cuneiform*. University of Texas Press.

Schulte, S. M. L. (2015). *Writing the history of mathematical notation: 1483–1700*. Docent Press.

Sergeyev, Y. D. (2022). Some paradoxes of infinity revisited. *Mediterranean Journal of Mathematics, 19*, 1–28.

Urton, G. (2010). *The social life of numbers: A Quechua ontology of numbers and philosophy of arithmetic*. University of Texas Press.

Zimansky, P. (1993). Review of Denise Schmandt-Besserat's *Before writing*, Volumes I and II. *Journal of Field Archaeology, 20*(4), 513–517.

Index

K. A. Overmann, *Cultural Number Systems*, Interdisciplinary Contributions to Archaeology, https://doi.org/10.1007/978-3-031-83383-0

D

www.ingramcontent.com/pod-product-compliance
Lightning Source LLC
Chambersburg PA
CBHW071608200726
48454CB00003B/198
* 9 7 8 3 0 3 1 8 3 3 8 2 3 *